OUR FORGOTTEN PAST . . .

We are embarking on a journey back through time to a momentus event 12,000 years ago. This event dramatically altered the makeup of our solar system. It has been ignored, misrepresented, or forgotten by most, though we glimpse it, like a phantom darting in the shadows, out of the corner of our eye. It haunts us every day in the reality we have created for ourselves, and in the ways we treat one another and all life on this planet. It haunts us in the terrible contradictions we come to accept on a daily basis and feel powerless to change. It haunts us because it *is* the source of how and why we came to be who we are today. . . .

WATERMARK

The Disaster That Changed the World and Humanity 12,000 Years Ago

JOSEPH CHRISTY-VITALE

PARAVIEW POCKET BOOKS
New York London Toronto Sydney

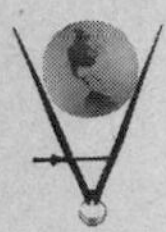

PARAVIEW
191 Seventh Avenue, New York, NY 10011

POCKET BOOKS, a division of Simon & Schuster, Inc.
1230 Avenue of the Americas, New York, NY 10020

Library of Congress Cataloging-in-Publication Data

Christy-Vitale, Joseph.
Watermark : the disaster that changed the world and humanity
12,000 years ago / Joseph Christy-Vitale.—1st Paraview Pocket
Books trade pbk. ed.
p. cm.
Includes bibliographical references and index.
ISBN 0-7434-9190-4 (pbk.)
1. Catastrophes (Geology) I. Title.

QE507.C48 2004
576.8'4—dc22 2003069037

First Paraview Pocket Books trade paperback edition June 2004

10 9 8 7 6 5 4 3 2 1

Manufactured in the United States of America

To all the heroes and teachers
past, present, and future.

CONTENTS

1. In the Beginning *1*
2. A Fantasy for Tomorrow *7*
3. The Tale of the Living and the Dead *13*
    The Living Evidence 13
    The Evidence of the Dead 21
    First Meditation 31
4. Phaeton, the Shining One *35*
5. Earth's Story *43*
    Upheaval 43
    Cosmic Assault 49
    Ice 52
    Scale 60
    Second Meditation 62
6. Phaeton: Into the Breach *65*

7. Our Ancient Ancestors, Part 1 73
Myth 79
Paradise 81
Language 83
Health 84
Spirit 88
Domestication 90
Medicine 93
8. Phaeton, the Planet Killer 99
9. Our Ancient Ancestors, Part 2 107
Energy 108
Architecture 112
Mathematics 115
Astronomy 119
The Flood 125
Third Meditation 128
10. Phaeton and Earth 131
11. After the Flood 147
The Heroes 147
The Teachers 157
Yesterday, Today, and Tomorrow 167
Fourth Meditation 177
Chapter Sources and Notes 181
Acknowledgments 205
Index 207

A book must be the axe for the frozen sea inside us.

—Franz Kafka

1

IN THE BEGINNING

Twelve thousand years ago countless humans, animals, and plants perished almost overnight, and great portions of the world were drastically and violently altered. Our ancestors, numb with shock and exhaustion, faced a challenge: either cease to exist as a species or survive in a grave new world. This is not exactly the history most of us were taught in school. Yet it is there, told in the scars on stones, the broken and buried bones of animals, and the memories of our species.

We thrust a stick into a clear pond, and it looks broken. It is just a distortion of light, but our eyes, ignorant of the physics behind it, see only a broken stick. As we look back in history, time distorts what we perceive, and even though we believe we understand its nature, we see history like that broken stick. This distortion is then magnified by both facts and our subjectivity, leaving us, in the end, with a flawed view of our past. This view, when enough people believe it to be true, becomes our worldview or paradigm.

This book is about a subject as simple yet as complex as that stick. It involves time, space, the Ice Age, Paradise, the nature of God, how we came to be who we are, and the end of the world. In other words: everything.

Our story begins in the late 18th and early 19th centuries when scholars and scientists developed a global paradigm. They had grown aware, through accumulated evidence, of immense scarred rocks scattered across the world in unexpected places. They also found deep deposits of sands, gravels, and mud in valleys and on mountaintops, broken and shattered animal bones in vast numbers, and made the first discoveries of frozen mammoth carcasses in the Arctic. From this hard evidence these scientists and scholars concluded that at some time in the not-too-distant past, the world suffered an appalling disaster. The hypothesis they developed was called Catastrophism. Much debate determined that water, in massive and swiftly moving amounts, was probably the main culprit. Among scientists and the general public many saw this as evidence of the Biblical Flood. To the surprise and pleasure of the clergy, science now supported their religious convictions. Yet even as they spoke of this momentous conversion from the pulpit, the Floodwaters, in some minds, were already beginning to freeze.

In 1830, Charles Lyell, a lawyer and amateur geologist, published his *Principles of Geology*. He insisted that instead of a sudden worldwide flood, a long, gradual accumulation of debris over millions of years had created the evidence. The hypothesis came to be called Uniformitarianism. He went on to say that the cause was terrestrial rather than cosmic and divine in nature, and if ever there was a catastrophic flood it was a regional and not global event. Lyell's idea proved to be popular among scientists. Over the following century his approach to geology and history replaced the catastrophic point-of-view and opened the door to the concept of the Ice Age, Darwin's revolution in evolution, and more recently the belief in continental drift and plate tectonics. Considered together they have established the current unifying paradigm of our world.

Today the wind has changed. Our current fascination with Uniformity and our belief in the long, slow geological and evolutionary process is beginning to crumble, like immense stone blocks falling from a walled city.

How do we know the walls are crumbling? Imagine a barking dog nipping at your heels, demanding your attention. Most of us would look down to see what the commotion was about, because from our earliest memories this companion has alerted us to things we should pay heed to. Our association with dogs has changed us profoundly, though some consider them a nuisance. Science would call this pesky canine behavior an anomaly. An anomaly is an unexplained fact, a fact that can contradict part of the prevailing paradigm. When enough anomalies are taken together, creating a hypothesis that better explains the nature of life, they can alter or push aside an old crumbling worldview.

This push has been occurring, for a number of years, with patience only Job or a glacier could muster. It has come from many quarters. Among others, are the observations of Immanuel Velikovsky, with his *Worlds in Collision* books, which have received either howls of indignation or the cold silence of an empty church from the orthodox science establishment. His ideas in general have been rejected but not refuted. Undeniable truths have survived and become seeds that have, over the years, erupted through to the public and are splintering and cracking the foundations of the current paradigm.

The first seed sprouted in 1972 when Niles Eldredge and Stephen Jay Gould proposed their "punctuated equilibrium" hypothesis. It says that a period of geological and/or evolutionary stability can be regularly interrupted by swift and even radical change. In other words, catastrophes can happen. A serious crack in Lyell's Uniformity then appeared in 1980 when Louis and Walter Alvarez introduced their impact hypothesis, which attempts to explain the mass extinction of the dinosaurs 65 million years ago. Their idea of a comet or asteroid colliding with Earth was first met with the same shrill reception as Velikovsky's ideas, but today most scientists accept it. Both "punctuated equilibrium" and the Alvarez impact have helped move science away from a narrow and extreme view of Uniformity toward a middle ground where occasional (in geological time) catastrophes occur.

The cracks are also appearing, rarely noticed by the general public or even science journals, down corridors and around coffee machines of universities, museums, and in the field. The debates revolve around the nature of these catastrophes and whether they ever occurred, particularly the one central to our story. The discussions are heated, because a new scientific revolution is being born. For some, change is difficult.

In this book we are embarking on a journey back through time to a momentous event 12,000 years ago. (Sources agree that the actual event described in these pages occurred 11,500 years ago, rather than the 12,000 year figure I use in this book. I've added 500 years—a blink of an eye in the geological time span—merely for the convenience of a nice round number, and for ease of discussion.) This event dramatically altered the makeup of our solar system. It has been ignored, misrepresented, and forgotten by most, though we glimpse it, like a phantom darting in the shadows, out of the corner of our eye. It haunts us every day in the reality we have created for ourselves, and in the ways we treat one another and all life on this planet. It haunts us in the terrible contradictions we come to accept on a daily basis and feel powerless to change. It haunts us because it *is* the source of how and why we came to be who we are today.

We will trace a path littered with evidence left for us by a trinity of testimony: biology, geology, and our human memory. Combining into an image beautiful and terrible, consistent and complete, it will challenge what we believe we know about our distant past. We will follow the drama from its beginning to its inevitable end, the desperate aftermath, and life's scarred resurrection on this third planet from the sun.

In a sense it is a pilgrimage, as we stop along our way to look upon anomalies, recall ancient memories, and realize that we are reclaiming our past. These moments may startle us, like a dog's sharp bark, and when we gaze up the world will begin to change. Not physically but intuitively. A pilgrimage is really a journey to oneself, and this book is no different. As we proceed, the complex and incomplete hypotheses of science and the blinders of religious beliefs will drop from our eyes. This journey goes beyond dogmas to look upon the past with clear eyes and to witness the spirit of all our ancient ancestors, as they stood on the edge of the abyss. We are going to remove the stick from the water, gaze upon it unbroken, and remember.

2

A FANTASY FOR TOMORROW

We will begin with something familiar: the end of the world. From our earliest myths to apocalyptic literature to motion pictures, such an event has been with us from our earliest days. We will discover its origin. But first imagine that tomorrow an announcement is made that an unidentified fiery object has appeared at the edge of our solar system. The report goes unnoticed by most, but anyone who has access to a telescope has it focused on that not-so-distant point in the sky. In a few days, on a clear night, its brightness can be seen with the naked eye. "No need for concern," officials announce, but one of the first thoughts that pass through many people's minds is, "Just like in the movies." This thought, however, soon disappears, leaving an unsettled feeling in its wake.

Over the following weeks the object's growing prominence in the night sky is evident, and it comes to consume more and more of the media's attention. People nervously begin to look up. Around the world many anxiously ponder this sparkling light, and scientists soon determine that this visitor is a piece of star matter. A distant star went supernova in a titanic explosion thousands of years ago and fragmented into millions of pieces. This fiery remnant is bisecting our solar system on its journey from that catastrophic event. Without a doubt, it is going to miss Earth. There will be no collision. We utter a global sigh and talk about this once-in-a-lifetime experience. Then a reversal; we are told that because of its mass and the closeness with which it will pass, the object will have an effect. How much is not clear. As the media begins promoting fantasies of disaster, numerous wits dub the object "Death Star." Governments struggle to keep their citizens calm. Over the next few months, this visitor passes unhindered through the outer planets' orbits, and soon moves through the asteroid belt, consuming thousands of smaller asteroids while capturing a dozen or so larger ones as satellites.

Now clearly visible during the day and brighter than a full moon at night, the first effects of the gravitational and electromagnetic attraction between Earth and this now unwelcome stranger become apparent. Like a human body fighting off a virus by creating a fever, Earth's temperature rises by the hour. The air soon crackles with electricity, winds heat up and blow with greater force, plants wither, and forests turn to dry tinder. Ocean tides around the world begin responding to the pull of Death Star's powerful gravitational force. Rumors race around the globe. Find shelter, some say. Flee to the mountains. Build an ark.

Below our planet's crust, unnoticed, an even larger sea of magma becomes more fluid as a result of the rising temperature in the Earth's core. Earthquakes increase worldwide as Death Star approaches. National power grids and communication networks fail. Water supplies to populated areas are cut off. Fires break out, and whipped on by hot, howling winds, rage unchecked through cities and forests. The oceans, drawn by the relentless pull of Death Star's gravity, move north and submerge coastal cities. Earth's surface begins to buckle, and frequent and powerful quakes shake the world. Civilization, overwhelmed by the magnitude of the growing disaster, collapses.

The planet's rotation slows, its winds roaring beyond hurricane force, its ground now continuously quaking. Large parts of northern Canada, Alaska, and Siberia sink beneath the rising oceans. Earth's crust, with volcanic force, ruptures into deep chasms, releasing enormous amounts of lava and poisonous gases. Cities, great rainforests, and mountains disappear in seconds. The sky darkens with dust and smoke. Earth and Death Star exchange immense bolts of electrical energy across space, as their electromagnetic fields come into full contact. The sound is shattering; the ground reverberates like a drum under each impact. Earth tips over and spins madly into darkness, then daylight as its axis slips. Some of Death Star's satellites are captured by our planet's gravity and break up, descending into the chaotic atmosphere, all aflame, raining down on the nightmare below.

Millions of animals, people, and plants perish. Landmasses collapse as the great magma seas below the crust are pulled to the north like the oceans above them. The world seems to be on fire except in the far north, where the mighty oceans, crushing the land below them, are massed mountains high, held in place by the uncompromising Death Star. Our sun's light can no longer cut through the dense layers of smoke, steam, dust, and gases that whirl and roar across the planet's changing face. As Death Star passes and moves away, its restraining influence weakens and the oceans above and the magma below begin to flow southward, slowly at first, then faster and faster toward the equator. This flood drowns the holocausts of fire and sweeps away everything except the tallest and mightiest mountains. The magma rises and falls in shuddering waves beneath the newly forming landmasses. Death Star passes beyond the solar system. Heavy dust clouds block out the heat of the sun, and temperatures plunge across a darkened and exhausted globe. Glacial conditions rapidly wrap the planet in an icy grip.

The humans that survive hang on by the slimmest of threads. Uncounted plant and animal species disappear forever. For decades it remains dark and cold, but eventually the clouds thin out, and the sun's heat begins to warm a deeply wounded world.

What remains of the vast global civilization that existed a few short years before? Just scattered groups of a new Stone Age people, on a world scorched, remolded, and washed clean, desperate to survive at any cost. What stories, learned from their parents, will the first children born after the disaster tell, when they are old, to their grandchildren? What tales will those grandchildren tell their grandchildren of that world-changing disaster, of that civilization that exists for them only in stories? What names and deeds will be remembered and how? And over the long years, then centuries, then millennia, what will be remembered of that distant time? When a new race matures, and its people look back into their past, what will they see? How will they interpret it? How will they write about it? Perhaps like this . . .

3

THE TALE OF THE LIVING AND THE DEAD

THE LIVING EVIDENCE

We take our first step, in remembering our forgotten past, by looking at our planet's biology. Animals and plants, both living and dead, appear in locations worldwide that raise difficult questions, pushing scientific disciplines into opposing camps, and contradicting the established view of our history. How did these plants and animals get to be where they are? Such a simple question is rarely asked, because when you do you enter a minefield filled with profound implications. But we *do ask,* and from these inquiries will emerge a picture of a world different both in physical appearance and biological diversity.

A V-shaped formation of cranes moves with sureness

through the silent air. Their senses, harmonized with the natural world engulfing them, respond to a yaw in the magnetic field and a shift in the wind, harbingers of a storm. This living geometric pattern turns ever so slightly and begins a long arced descent to the peaks now rising from the ocean on the distant horizon. Those mountain peaks we call the Canary Islands lie off the Atlantic coast of northwest Africa. They are volcanic islands partially created by massive volcanic eruptions 12,000 years ago. They were known to the Greeks and named by the Romans over 2,000 years ago. Looking beneath the surface of these islands, we find burrowing into the soil an earthworm.

All earthworms are blind and deaf and are unable to survive immersion in salt water. Our Canary Island earthworm, oligochete, is a species so similar to earthworms in southern Europe that zoologists suggest these beautiful islands were once connected to the European continent. Orthodox science acknowledges the volcanic origins of the islands, disagrees with the land connection theory, and ignores the earthworms. If the Canaries were never previously attached to a landmass and were volcanic in origin, how did oligochetes come to be there? Two ways suggest themselves: a special earthworm creation on the Canary Islands, or they were brought there by some unknown seafaring Johnny Appleseed who took the time to distribute them on each individual island. More likely, however, these islands *were* connected with Europe and Africa. Because oligochete's similarities are greater than their differences from their European kin, many zoologists are convinced this once-upon-a-time land connection, which today rests beneath 10,000 feet of ocean water, must have existed in the not so distant geological past.

Crossing the South Atlantic reveals another mystery: the monk seal. Rarely venturing into open ocean, this endangered animal is found along the eastern coast of South America and, until the last century, in the West Indies. Back across the Atlantic monk seals can be found barking on the shores of West Africa and the western Mediterranean. How did these seals get from West Africa to South America without crossing the open ocean? Any idea concerning mysterious seafarers transporting them across the Atlantic can be dismissed. According to the current continental drift hypothesis, the American continents became separated from the old world 65 to over 130 million years ago. At that point the carnivorous, arboreal, tropical forest-dwelling ancestor of the monk seal was not even a glimmer in Nature's eye. Another cause is needed.

Below Earth's oceans there exist 5,000 species of sponges, but only 20 are native to freshwater; these are unable to survive in salt water. One, *Heteromeyenia ryderi,* is found only along rivers and lakes from Florida to Newfoundland along the Atlantic coast of North America, and across 3,000 miles of salty ocean on the west coasts of Ireland and Scotland. This distribution has been a puzzle to those scholars who are not worried about what questions to ask, or what answers may arise.

Appalachia is a name that many of us associate with the mountain range in the eastern United States, but geologists and botanists also give it to a continental-size landmass that they believe once connected, via Greenland, both Europe and North America. Most of this ancient land, they also believe, sank two miles below the surface of the sea at the end of the Ice Age, 12,000 years ago. Prior to that, the landmass provided, among other things, the freshwater sponges, like *Heteromeyenia ryderi,* the opportunity to find their way to their present habitats. While the Ice Age conjures up images of massive walls of ice entombing land and sea for thousands to millions of years, the fact that our sponges require freshwater transportation and a temperate climate comes into direct conflict with such images.

There was once a great body of water known to geologists as the Miocene Ocean. Its waves brushed shores from what is today central Europe to as far as Lake Baikal in Siberia, 6,000 miles away, and from the Black and Caspian Seas up to the Arctic. Its disappearance must have occurred geologically recently because the seal populations that live today in Lake Baikal and survived well into the 1800s in the Caspian Sea over 2,000 arid miles to the west were identical.

Almost 3,000 miles southeast of Africa lays Kerguelen Island. Europeans discovered it in the late 1700s, and it was first investigated in the early 1840s when English botanist J. D. Hooker and the Antarctic exploration ships *Erebus* and *Terror* stopped by on their way back to England. These isolated mountaintops turned out to be the home to freshwater fish that are identical to species in New Zealand (over 6,000 miles to the east), and South America (6,500 miles to the west). Kerguelen Island and Heard Island, another 400 miles farther south, are also home to a rare wingless fly. Hooker believed these islands were once a part of a large southern continent now vanished, as this was the only way he could explain the animal and plant life.

Spread across the vast stretches of the Pacific are thousands of isolated islands. Widespread throughout these islands are land shells, from the mollusk family *Clausiliacea*. They have a natural aversion to salt water, and their distribution and close relationship has convinced a number of botanists that a large continental landmass once existed in this vast oceanic region. What happened to that landmass is a mystery.

These are only a few of many examples worldwide. From seemingly insignificant earthworms to insects, crustaceans, mollusks, amphibians, reptiles, and mammals, we find all these living creatures and their closest relations sometimes located thousands of miles apart, insisting by their existence that our planet's geography was remarkably different than we know it today. The idea of being separated by the slow, gradual process of continental drift is called into question because they are too closely related. The evidence creates the impression all these animals were stranded by some sudden and dramatic event in the recent geologic past.

• • •

Just as animals are found where they should not be, plants also have some interesting stories to tell. While it is true that wind, sea currents, birds, or man can carry the seeds of plants to remote places, the following examples have challenged those distribution methods and forced scientists to look for other explanations.

To the chagrin of other disciplines of science, many botanists long acknowledged that identical plants found in Europe and eastern North America lead to a view that a land connection existed in the not too distant past. The drifting of continents millions of years ago cannot account for their current distribution. There are freshwater weeds, predating the arrival of Europeans, which are found both in the British Isles and North America that could have only been spread by a land route. For example, the freshwater plant species *Najas flexilis* is found in North America but also in parts of Ireland, the Isle of Skye in Scotland, and Cumbria in England.

Another interesting example is the tree Eugenia. It is presently found on Mauritius, off the coast of Madagascar, and on Marion Island, just north of the limits of Antarctic icebergs in the southern Indian Ocean, as well as on the Solomon Islands in the distant Pacific. In these cases the plant's seeds are not used by birds for food (practically eliminating their transportation by air over great distances), nor can they survive immersion in salt water for the periods necessary to float from one island or continent to the next. That leaves wind and man. Transport by wind has been ruled out because of the vast distances involved, air currents blowing the wrong directions, and the fantastic odds of a seed landing in a suitable location. The Johnny Appleseed distribution network would have had to occur in such distant ages that our current view of ancient man and his ability to travel globally would have to radically change.

Back on Kerguelen Island, Hooker observed that the cabbage found there is the major food source for a short-necked duck called a teal. The plant's seed, however, is too soft and easily destroyed to pass through the duck unharmed. Because of this fragility, it is not plausible that the seeds would be carried any distance with success if somehow caught on the animal. Yet the same cabbage is also found on the McDonald group of islands (Heard Island is among these) and Marion Island. These islands are scattered over 3,000 miles of ocean. Kerguelen Island also has flora similar to that found on New Zealand (4,000 miles east) and South Georgia Island (nearly 5,000 miles to the west). Science finds this difficult to explain, but recall that Hooker was an early advocate for a lost southern continent.

The volcanic island Tristan de Cunha, halfway between South America and Southern Africa, has plants common to both areas. Botanists have ruled out wind and animals as methods of transportation, so some scientists have suggested that icebergs during the Ice Age carried them. One can only wonder how icebergs calving in the midst of an ice age found the seeds in the first place and then, like a cruise ship, sailed them to a friendly island port.

It is unusual to find more than three or four species of the same genus on oceanic islands (meaning islands that supposedly have never been connected to a continent). Their isolation prevents development of an abundance of species. However, the oceanic Canary Islands, where we began our journey, have from three to ten times that number of species. The Hawaiian Islands are another instance where the flora is too rich to have developed just on those islands. The idea that Hawaii was once a part of a much larger landmass has been suggested by science and remembered as such in myths. The fact that many of the Hawaiian plants have mainland American relatives, 2,500 ocean miles away, compounds the mystery.

In the current paradigm, the idea of continental drift insists that many landmasses we have mentioned, Appalachia and Hooker's lost southern continent, never existed. Yet continental drift is just a hypothesis, and the living plants and animals, as we have seen, as well as various methods of alternative transport, challenge that belief. The evidence suggests that in our recent past there *were* vast landmasses and continents that we have since forgotten. Something happened to radically change it all.

THE EVIDENCE OF THE DEAD

We turn now from the living evidence to that of the dead. This testimony will support the revised view and raise even more questions.

Lignite, sometimes called brown coal, is found worldwide. Similar to soft coal and peat, it is used as a fuel in northern climates. In lignite you will usually find the texture of the original wood that formed it. Because of this, its age is not considered great. Difficult questions arise, however, when the trees found in the lignite are not local. Where do they come from, and how did they get there? In Germany many of the lignite beds are associated with swamp bogs. But the remains found in the bogs are sequoia, a tree that does not grow in swamps. They could only have been carried and buried here. Australia has similar deposits. Near Morwell, in Victoria, the lignite is found 200 feet below the surface and is nearly 800 feet thick. The numbers of trees found are prodigious and well preserved and include many that are not common to the region. Again, they could only have been carried and *then* buried there.

In many places lignite is found in what are called drift deposits. Sands, gravels, and plant remains make up this densely packed "drift" that fills valleys, covers mountain flanks, and accumulates to depths of thousands of feet. Standard thought explains these drifts as the debris of the great retreating ice sheets, even though they are found all around the world, and in areas that have never known an ice age. That much of these "drift" remains—leaves, twigs, berries, nuts, and moss—were not crushed by the ice sheets is ignored.

In Brazil, in Cameroon in Africa, and on the Indonesian island of Java, lignite beds are found with fungi and algae attached to leaves whose fibers are still green with chlorophyll. For this to occur burial must have been swift—*not* a gradual event. These beds contain not only leaves local to those areas but from around the world as well.

In the United States, at locations throughout Wisconsin and south to Iowa, then east through Illinois, Indiana, and Ohio there are areas where remains of immense forests of conifers and deciduous trees lie buried beneath the soil. They were first discovered in the 1800s when people began sinking wells into the ground only to find, from a few feet to 30 feet down, these ancient dead forests. Not having yet petrified into stone, they are believed to be geologically young. Since their discovery, this anomaly has perplexed scholars. The areas where they are found are not volcanic, and if some prehistoric flood, which has been suggested, buried them it would have to have been of tremendous size.

Southwest of James Bay in Ontario, Canada, there are lignite beds that descend to depths of 125 feet. In the Northwest Territories, including much of the barren land east of the Canadian Rockies, the same anomaly is present. Up above the Arctic Circle, where even in summer freezing winds blow, are the Queen Elizabeth Islands. Scattered across these desolate, treeless lands, usually believed to have been buried under immense ice sheets, enormous quantities of "drift" are found—complete with leaves, pinecones, and acorns—up to 300 feet *above* sea level and in jumbled masses sometimes 40 feet in height. While mining for gold in Alaska's Goldstream Valley, miners uncovered thousands of trees, in their original positions, sheared off six feet above their bases and then buried. How it all came to be there is a mystery.

In treeless Greenland, the remains of pinecones, acorns, sequoias, oaks, maples, and magnolia are found together in confused masses in a land today synonymous with ice, *not* temperate forest. On Iceland and the high arctic Spitsbergen islands, deposits also suggest a much more temperate climate in the recent geologic past. It has also puzzled scientists that there are no transitional plants found in these areas. Whatever it was that caused it, the change from temperate to arctic conditions was abrupt. In all cases the flora are mixed together in a violent and chaotic burial.

Prehistorians gave the name Fennoscandia to a landmass that, 12,000 years ago, sank into what is today the Arctic Ocean. It included much of northern Eurasia and the now numerous island archipelagos that are found in the Arctic Ocean, some lying less than 500 miles from the North Pole. The remains of immense forests here staggered the first scientists who explored these lands in the 1800s. Twelve thousand years ago forests of alder, elm, and oak reached far into the Arctic from the Ural Mountains and Siberia. Today they lie buried in frozen ground where no bushes and trees can survive.

On Kotelnyy Island, dead trees stand where they once grew thousands of years ago; their tops are twisted and smashed, or they lie in confused mounds of timber up to 180 feet high. Leaves and cones are found among the remains of once huge forests that cover the shoreline for miles on the New Siberian Islands. Along the northern Siberian coast, these relics of a forgotten time stretch for hundreds of miles. The people who lived in this inhospitable land over 100 years ago acknowledged differences in the "driftwood" they used for fuel. One was called "Adam's wood" for the most ancient of the dead trees, while the other was known as "Noah's wood." This was the wood brought to their land by the Flood.

According to the Ice Age hypothesis, most of this land was buried under immense ice sheets for thousands to millions of years. These forests should not even exist. But they are there standing like forgotten skeletons in the far north. Something rapid, powerful, and global in its power destroyed them, leaving everything in chaotic ruin.

The loss of animal life 12,000 years ago was also profound. Biologists have difficulty communicating the magnitude of extinction that occurred during this time period. In *When the Earth Nearly Died,* paleogeographic researcher Derek Allan and geological surveyor J. Bernard Delair summarize this frustration when they write "immense herds of diverse animals utterly vanished off the face of the Earth for no obvious biological reason. They were seemingly virile, numerically strong faunal groups, well adapted to their natural environment, yet, geologically speaking, they disappeared with frightening abruptness." Scientists who do not believe that a catastrophe occurred so recently in our geologic past invoke the term "climate change" to explain away the deaths. The climate changed and various immense herds perished separately during the numerous advances and retreats of the ice sheets. But, as we shall see, this explanation ignores or misinterprets the existence of the evidence.

In Europe, bones mixed with plant debris are found in many lignite beds and drift deposits. Near Lake Zurich, in Switzerland, lignite beds have been studied that contained bones of extinct mammoths with hippopotamus, rhinoceros, and other contemporary animals mixed haphazardly with vegetation. Deposits in Germany find lignite impressed with tropical plant, animal, and insect remains. Most of the insects are considered modern fauna, identical to insects found today. This indicates a recent geological demise and acknowledges a very different climate than Germany currently enjoys. Sites in England have been discovered where buried forests of gigantic trees, some still with their berries and nuts, are mixed in with the bones of mammoths, horses, hippopotamuses, and other species.

The arctic lands of Alaska and Canada have debris fields

called "muck beds" that tell of wholesale slaughter. These fields are widespread and can be up to 200 feet thick. Many of the muck beds in Alaska are located in regions *that have never been glaciated*. They are all crammed with plant matter, shattered remains of trees, boulders, wedges of ice, up to four different layers of volcanic ash, and great numbers of animal bones. A partial list of bones found, including currently living and extinct species, reveals the diversity: squirrel, beaver, sheep, wolf, antelope, caribou, saber-toothed cat, lion, horse, camel, bison, musk ox, mastodon, and mammoth. In many instances parts of their body tissue, skin, and hair were preserved. But these are not complete skeletons. They are broken and crushed remains, as if their bodies were first torn to pieces and then buried and frozen in these terrible mass graveyards.

On Spitsbergen, "drift" is found almost 30 feet above sea level woven with whalebones. More sites with whale and other marine mammal remains have been uncovered along the ridge tops of the Santa Paula Mountains, north of Los Angeles. At one site the remains of a large seal were discovered 2,000 feet above sea level. Also near Los Angeles are sites where both land and sea species, usually believed not to have existed in the same time period, are found buried together in a dismembered and helterskelter state.

Throughout the Mediterranean region, there are caves and fissures filled from the narrowest cracks in the walls and floors all the way up to the roofs with the smashed remains of hundreds of species of animals, reptiles, amphibians, crustaceans, and sediment. From the Rock of Gibraltar to Sicily, Malta, and Crete scientists have uncovered these dens of death. Sea and land shells, turtle, shark, lizard, hyena, bear, elephant, hippopotamus—including pigmy hippopotamus found living today only in West Africa—bird, rodent, deer, ox, lion, woolly rhinoceros, and mammoth have been found crushed together in these caves, fissures, and crevices.

Sometimes called bone-caves, these final resting places for plants and animals are not exclusive to the Mediterranean. They are found throughout the world. Animals are buried together regardless of age, sex, size, geographical distribution, and compatibility of species. Sometimes their bones are still attached to sinews, muscles, and flesh. The United Kingdom has numerous caves resembling charnel houses. Oreston, Brixham, Kent's, and Victoria are just a few caves where incredible numbers of incompatible and diverse animal bones are entombed. In Vallonet cave, near the French-Italian border, lion, monkey, elephant, rhinoceroses, and whale remains have been found. In Australia, at Wellington, and Borel, we find caves filled with these chaotic burials to depths of 200 feet. Caves in Israel, Lebanon, and Syria have revealed woolly rhinoceroses and other animals considered "northern" mixed together with "southern" animals such as the water buffalo.

Near Beijing, China, a group of bone-caves and fissures at Choukoutien led to the discovery of Ice Age animals buried below the remains of "earlier" animals. None were ancient enough to be fossilized and so are left unexplained by scholars who follow the current paradigm. Rodent, hyena, fox and wolf, saber-toothed cat, lions, tigers and bears, woolly rhinoceros, sheep, horse, camel, baboon, ostrich, and tortoise are animals from diverse locations and climates who shared in these caves a final violent burial.

In one cave at Choukoutien, mixed in with this plethora of animal bones, were shattered human remains: a European, a Melanesian, and an Eskimo.

South American bone-caves reveal staggering quantities of remains, especially in Brazil. One cave, near the Lagoa do Sumidouro, revealed, in addition to the usual wide variety of animals, the bones of over 50 humans. Men and women, infants and the aged were all buried in bone-shattering, brutal confusion.

Nor is North America exempt. Maryland has a bone-cave near Cumberland with lemming, mink, badger, elk, crocodile, and tapir. Again we see both "northern" and "southern" animals. Some suggest that humans and carnivores used the caves as dens. But they cannot account for the sheer numbers of animals buried in these caves. At one time scholars suggested that it was the movement of ice, both advancing and retreating, that trapped these regionally different animals and humans in caves. The thought of mammoth, monkey, weasel, and whale being chased into and trapped in the same caves by an ice sheet is certainly picturesque. Fortunately that "scientific" theory has been abandoned. It has been acknowledged that, whatever the cause, most of the caves filled quickly from floor to roof.

The La Brea Tar Pits in Los Angeles are another interesting case. The tar pits are, in fact, not pits at all. They were springs, pools, ponds, and lakes. Scientists dug the "pits" in their excavations. Similar to the muck beds in the Arctic, the "pits" have been viewed as natural traps, which animals would wander into, become stuck in, and eventually die. They may have called out, attracting others of their kind who also became trapped, and perhaps carnivores as well who came looking for a quick lunch but fell victim themselves. Following that view, you would expect to find complete skeletons in the beds. But this is not the case. Almost all of the enormous amounts of remains that have been excavated from La Brea have been shattered, smashed, and mixed in chaotic fashion. These were not some animals stumbling into a tar pit, but thousands that were torn apart violently, *then* buried. The remains are around 12,000 years old.

Back in the north, Siberia is well known for its frozen mammoth carcasses. These are still being discovered, and while few in number, speculation and hypothesis whirl, like a blizzard, around them. Some scientists believe numerous, gradual climate changes melted the ice sheets and broke up the land, creating the crevasses that these mammoths fell into, where they died and were frozen. Some think the climate changes produced occasional meltdowns, at various times over tens of thousands of years, which sent water bursting through ice dams and caught the animals unaware. In either case the causes remain comfortably within the Ice Age hypothesis. Yet some botanical evidence creates a paradox for science. The climate, shown in floral studies, was considerably milder than today. These mammoth carcasses were not found in ice, but buried in now frozen sediment or mud, meaning the ground was *not* frozen when the animals died, otherwise they could not have been buried whole. The ground then had to freeze quickly, not gradually, before the animals decomposed.

Supporting this scenario, across the Siberian tundra many sites reveal an identical story of great destruction of freshwater plants, willow, alder, and other tree species, as well as mammoth, wooly rhinoceros, bison, horse, and countless additional animals. These are not carcasses but bones. All are broken, buried, and frozen, sometimes in mounds like giant anthills scattered across the horizon or like flotsam left behind on some vast seashore. Russian scientists evaluating the mammoth bones found in an area of 527,000 square miles, or an area a little smaller than Alaska, estimate that there were over 11 million mammoths. It has also been suggested that by adding an equal number of wooly rhinoceros, even larger numbers of bison and horse (whose bones are consistently found in larger numbers than the mammoths) that the number of animals that inhabited that piece of land reached upwards of 174 million.

The question must be asked, where did the food and

water come from to sustain these creatures? If there was an Ice Age 12,000 years ago, with an annual six months of darkness, the long dry winters would leech out whatever nourishment any plants had and freeze freshwater supplies. The animals could eat all the moss and snow they could find (and the larger ones would need hundreds of pounds per day), but they would die, due to lack of protein and energy long before winter ended. Add to those 174 million animals the hundreds of millions more in the rest of Siberia, Northern Europe, Greenland, Canada, and Alaska, and you begin to understand that such immense herds would not have existed if Ice Age conditions, as described by current scientific opinions, were real. Those animals buried together with the uncountable broken forests in Earth's arctic regions contradict in the strongest possible way the orthodox view. Not ice or a slow, gradual process, but a single catastrophic event is the only explanation for the uniform nature of this holocaust.

FIRST MEDITATION

As the sun begins to dissipate the morning mist, we pause in our journey and enter into an assembly filled with people like ourselves. A short distance away pine trees can be seen, and chrysanthemums blossom beside a pond. Cranes are standing in the water, gazing at us, suspicious of our arrival. This is a place of remembrance, of testimony. We sit among the rows of public benches and recall what the plants and animals, both living and dead, have told us. The living organisms have no reason to be where they are according to the current scientific paradigm. Yet they are there. The immense remains of the dead tell us that 12,000 years ago an overwhelming disaster arbitrarily and completely slaughtered countless life-forms. They are all anomalies and disparate and until now isolated enough to not present a threat to conventional science. But they are barking dogs, nipping at the heels of science's dragging feet, and their testimony can no longer be denied.

Our modern error, which is repeated in countless publications, is the attempt to remember the distant past on *our* terms. We do this with the impiety of a race with a short memory and shorter life span. This impiety is like the passion of a tyrant who with closed eyes demands that only his story be told. We need to be witnesses, and with open eyes see the entire picture.

The evidence we have discussed so far begins to seriously question whether there was an Ice Age at all. If so, it cannot have been as extensive as we are led to believe. Our planet, 12,000 years ago, was more temperate, humid, and lush than today. Vast grasslands and immense cathedral-like forests of sequoia, cedar, oak, and thousands of other species blanketed the globe. Together with tremendous expanses of ferns, these trees blessed Earth with a richness not seen since. She was also inhabited by vast herds of animals on landmasses shaped differently than they are today. The biological evidence suggests a series of east-west configurations instead of the north-south continental arrangements we have today. Then something happened. Immensely powerful and appalling in its completeness, it operated on a worldwide scale. Entire species were exterminated, others sundered forever from their own kind by new geographical conditions. The Ice Age and continental drift with their slow, gradual processes do not allow such an event. Another explanation is required.

4

PHAETON, THE SHINING ONE

If you are reading this in the evening, take a moment to step outside and gaze at the night sky. Like a colorful medieval tapestry, the sparkling glitter of stars and planets all journey tirelessly across this silent panorama. On a night like any other 12,000 years ago the announcement came: It was the beginning of the end. For uncounted ages prior to that, ten planets did their circle dance around the sun. All had axes whose inclination was minimal: North was at the top. There were no ringed planets. What moons there were orbited their mother worlds in uniform fashion. Asteroids and comets were extremely rare. Prior to that night it had been a very long time since anything happened in the House of the Sun.

On Earth stargazers in the Southern Hemisphere had already noticed a small point of light moving against the background of stars. Hebrew traditions recall that the Pleiades, the brightest star cluster in the sky, and the constellation Orion, both much lower in the heavens 12,000 years ago, were the region where the intruder first appeared. The Roman Ovid in his *Metamorphoses* mentions how it "climbed the steep ascent" into history. That this object was not from our solar system is understood in the Babylonian epic *Enuma Elish,* where its birth occurred in the "heart of the Deep," also called the Great Abyss. Psalm 19 states, "from the end of heaven he emanates."

History has many names for it, most were titles of powerful gods or monsters: horned, multieyed and many-faced. All took part in a war in heaven. Our Judaic Christian *Satan* is one, who when cast down from heaven covered the world with a flood that belched from his mouth. Nearly identical is ancient Egypt's *Set,* a serpent from darkness and chaos, the antithesis to peace, harmony, and order. From India, *Shiva,* terrible to behold, was also known as *Hara,* the universe destroyer. Mexico's *Tezcatlipoca,* the Aztec lord of wind, rain, and winter, was cast down from heaven. *Susano-wo* from Japan was a destroyer, a lord of darkness and whirlwinds. He too was cast down from heaven. The Greeks knew it as *Typhon,* a monster that vomited flame, lightning, and great wind, and from him comes our word typhoon. In ancient Mesopotamia its name was *Marduk,* an avenging sun and lord of thunder. Terrible winds and flames roared from his mouth. He was a creator of all things anew. *Tishtrya,* a star "that attacked planets" in Persian history, drowned humanity with a great flood. And to the Romans it was *Phaeton,* a starlike object that, in ancient days, rampaged across heaven and rained great destruction upon Earth.

Phaeton. Literally meaning "The Shining One," the word was borrowed from the Greeks by the Roman Ovid 2,000 years ago for his epic *Metamorphoses.* What was Phaeton, known by so many names around the world, which left such an indelible scar on our collective memory?

We know what it was not. Phaeton was not an asteroid or a meteor. Those objects' origins remain mysterious, with scientists debating whether they were once parts of a larger body or bodies that met with some catastrophic disaster or the remains of comets. Heavily cratered, asteroids number perhaps into the billions and range from a few feet in length to over 600 miles long. Many travel in family groups on bizarre orbits through our planetary system. How they acquired these scars and irregular paths around the sun is also a mystery. Though destructive, large asteroids (recall that the dinosaurs may have met their fate with one) and the much smaller meteors lack the mass needed to create a disaster on the scale described here. Also, like a mountain on a moonless night, they can be invisible to Earth-bound viewers until almost the last moment.

Phaeton was not a planet. Planets certainly have the size and mass necessary to trigger a worldwide calamity. However, the proximity of a larger planet would have spelled Earth's destruction, due to Roche's Law, which defines the minimum distance planetary bodies can approach one another without the smaller object being torn apart by internal tidal forces and/or external electromagnetic forces. The rings around Jupiter, Saturn, Uranus, and Neptune are within the limits of Roche's Law, and are most likely the remains of objects that approached too close to those giant planets.

A smaller planet nearing Earth would be expelled by electromagnetic energy bolts coming from Earth and/or torn apart by its own internal forces. But if the object were, as Allan and Delair suggest, "close to either Earth's mass and density, or of smaller mass and greater density," Earth's inhabitants would have faced a dangerous situation.

Such an object, moving at tremendous speed through a portion of the Roche zone, would not need to collide with Earth to create a disaster. And though large planets could be seen from a distance, they are not aflame, and mythic memories are specific about calling Phaeton starlike or a "ball of fire," not a planet.

Comets have been the most popular interpretation of this celestial object (they along with meteors and asteroids, are the main suspects in the demise of the dinosaurs). Where comets come from, how they are created, and how many there are out there is another mystery. Composed chiefly of ice and dust, the comets we are familiar with throughout history have not been of sufficient mass or density to induce a planet-wide disaster of the magnitude under consideration. Nor can they be seen by the naked eye—and thereby mark our psyche—until they are close to Earth. Our ancient ancestors are quite clear: What approached was as dramatic as it was unfamiliar.

Looking beyond our solar system we find another object worthy of our attention. Within the last 15,000 years our immediate galactic neighborhood has witnessed five supernovae explosions. One, the star Vela, is believed to have self-destructed between 11,000 and 14,000 years ago at a distance of 45 light-years from Earth. In cosmic standards, that is just around the corner. A piece of flaming star matter, not much smaller than the planet Earth, rocketed in our direction and, traveling anywhere between one-fiftieth and one-hundredth the speed of light, could have entered the outer limits of our solar system less than a thousand years after the explosion. The event we are remembering occurred around 12,000 years ago and, falling within the dates for Vela's fatal convulsion, suggests this piece of astral shrapnel was the source of the local disaster.

Vela's titanic detonation sent Phaeton and millions of astral companions, a solid wall of gas and fire, into space. The companions quickly lost that massed density, and like deadly pellets from a shotgun blast, spread their luminous destruction in all directions.

Forty-five light-years away, inhabitants of Earth witnessed the Vela supernova event. Our ancient ancestors first saw it as a brilliant flash in the day and night sky and later searched amid the stars for a point of light that could no longer be found. How well they understood what they saw remains unknown, but they and their children slept peacefully under that missing star for many years. Only generations later when the "Shining One" announced its arrival did the enormous implications reveal themselves.

As Phaeton and Vela's other unforgiving remnants raced to their destinies, they did not travel in a void. Floating ions, atoms, and electrons fill space. A fragment of an exploded star, probably with an electrical or magnetic field (a parting gift from its now deceased parent), would attract, as it sped through the galaxy at high speed, those ions, atoms, and electrons. Phaeton, gorging itself on this electromagnetic feast over hundreds of years of interplanetary travel, became a miniature sun.

As Phaeton shot through the doorway of our solar system, the gravitational and magnetic fields of the outer planets challenged it. Phaeton also came face-to-face with our tenth planet. As this unwelcome guest and forgotten planet neared one another, their magnetic fields reacted and traded energized lightning bolts, each object attempting to discourage the other from approaching too close. Traveling until now on its own path, Phaeton tried to resist the stressful pull of the giant planets. Their combined attraction caused the temperature within this child of Vela to increase to a dangerous level. The struggle continued, as the tenth planet and Phaeton closed in on one another, until it triggered a terrible pronouncement from the little sun. Phaeton exploded.

On Earth, billions of miles away, everyone looked up.

5

EARTH'S STORY

UPHEAVAL

We turn now to the tale Earth itself has to tell. The majestic European mountains known as the Alps are geologically very young. Swiss geologists believe that the Alps were no more than low hills 12,000 years ago. How far up one has to climb for a hill to become a mountain can vary by location, so let us call anything under 3,000 feet a hill. The Alps of 12,000 years ago were then no more than 3,000 feet high. Now compare those heights with the peaks that we see today: the Matterhorn measures 14,691 feet, and Mont Blanc 15,781 feet. Little math is required to understand how high these "hills" were uplifted and that it was not a long, slow, uniform process. Immense walls of stone, thousands of feet thick and up to 40 miles wide, stretching beyond the horizon, violently thrust up from Earth's surface. Like a gigantic ship plowing through the sea, it sent earthquakes roaring off its wake. The stupendous friction created deafening sound, and heat waves blew from the vast furnace-hot walls igniting the landscape. In this uplift the entire range of mountains slid north a hundred miles, in places grinding itself over younger rock. What would trigger such a catastrophic event?

In North America, from Yellow Head Pass in Alberta to Glacier National Park in Montana, a distance of more than 500 miles, the Rocky Mountains moved many miles in a lateral direction as they were thrust up 12,000 years ago. The Coastal Range and Sierra Nevada Mountains in California and the Cascade Range stretching from Oregon north to British Columbia also underwent rapid upheaval, sometimes up to 6,500 feet, in our recent geological past. In Southern California, the Coastal Range also reveals the remains of marine life doomed when they where uplifted more than 1,000 feet from the floor of the ocean.

Evidence suggests that the Andes, in South America, though an ancient mountain range, reached its great heights only 12,000 years ago. Scientists believe this upheaval to have been swift and accompanied by tremendous volcanic eruptions. One can travel to 11,000 feet to Lake Titicaca, on the Bolivian-Peruvian border, and still have to look *up* to see terraces of marine shells that were thrust up to 13,000 feet from the floor of a sea now almost 200 miles away. The innumerable ancient agricultural terraces scattered between 15,000 and 18,000 feet and the immense salt beds covering thousands of square miles, similar in composition to the salt in the ocean, demand an explanation other than the "slow, gradual process" we are led to believe occurred.

Continuing to Asia, we see more dramatic confirmation of planetary upheaval. The Tibetan Plateau, an area of more than 2 million square miles was uplifted more than 9,000 feet. The estimated date for this event, according to geologists, ranges from 12,000 to 15,000 years ago. Himalayan peaks show extreme youth that confound and challenge science. Indian scientists believe that the last savage thrust must have been close to 6,000 feet in elevation in parts of Kashmir. Here are found 12,000-year-old marine beds lying more than 5,000 feet above an ocean that no longer exists.

Upheaval is evident throughout the mountain ranges of Pakistan, Afghanistan, and Central Asia. Here within the recorded memory of ancient China lay a large sea known as the Great Han Hai. It extended more than 2,000 miles east to west and 700 miles north to south. It may have even been part of the Miocene Ocean mentioned earlier. Twelve thousand years ago, it existed 3,000 feet lower than it does today. It is believed to have been uplifted at that time with the Tibetan Plateau. As it rose toward the sky, the Great Han Hai's incredible volume of water poured over the rising cliffs as if from an overflowing bathtub.

When the Tibetan Plateau and the Himalayan Mountains were thrust up, the area immediately to the south collapsed. Called the Gangetic Trough, this land sank more than 6,500 feet. It has a width of 250 miles and stretches 1,200 miles along the now towering mountains. This huge trough is filled with debris from the so-called Ice Age.

Off North America's eastern seaboard, fishermen have brought back bones of mastodons, horses, musk ox, and more. The bones came from depths of 400 feet and from almost 200 miles offshore. They are believed to be 10,000 to 12,000 years old. It was here the lost land of Appalachia existed 12,000 years ago.

Cut into and scarring the Pacific Ocean's sea bottom are undersea trenches as deep as 35,000 feet and thousands of miles long. Parallel to South America's Andes Mountains is the undersea Chile-Peru Trench, which in places is thousands of feet more below sea level than the Andes are above sea level. If such trenches had been created through a slow, gradual process, these great slashes into Earth's crust would be filled with millions of years of sedimentation. Yet these submerged trenches show little sedimentary filling, suggesting that they are quite young.

Together with these colossal upheavals and collapses, volcanism occurred on an unprecedented scale. Volcanologists have charted upwards of 10,000 volcanoes that were erupting simultaneously some 12,000 years ago, spewing lava in enormous amounts across Earth's surface. Two examples: In India, throughout the vast Deccan Plateau, lava flows are several thousand feet thick and cover more than

250,000 square miles. The entire northwest of the United States, from the Rocky Mountains to the Pacific Coast, lies atop 12,000-year-old lava flows. In Seven Devils Canyon, the Snake River has cut through 3,000 feet and has yet to reach the bottom of the lava. The Columbia Plateau sits above lava that is believed to be 5,000 feet deep.

In the Atlantic Ocean we know that both the Canary and Azores Islands are volcanic in origin and were subjected to extensive volcanic activity 12,000 years ago. This resulted in the sea floors around the islands, particularly the Azores, being covered with lava. Called tachylite, this type of lava will disintegrate in seawater within 15,000 thousand years. These flows are considered 12,000 years old.

Recalling that there are biological connections between the Canary Islands (and the Azores Islands as well) and the European and African continents with plant and animal species, we need to consider how two deep-sea core samples, taken near the Canary Islands (from 10,500 to 18,440 feet), were found to be made up of very young *beach* sand. Only one conclusion can be drawn: 12,000 years ago this land was above the sea.

A mammoth crack traverses our Earth's crust. More than 40,000 miles long, it wanders both above and below the world's oceans. Part of it includes the Mid-Atlantic Ridge that stretches the entire length, north to south, of the Atlantic Ocean. It is also believed to be youthful because in many places the sedimentation is quite thin and in places nonexistent. On land this crack is called the Great Rift Valley. Formed around 12,000 years ago, it extends from Jordan in the Middle East southward through East Africa to Mozambique. In places the plateau above the Rift Valley rises 2,000 to 3,000 feet. Other escarpments plunge 9,000 feet into the valley below. The lakes along the Great Rift are very deep. Lake Tanganyika, for example, is the second deepest lake in the world, at a depth of 4,700 feet. All these Rift lakes are fjordlike, that is narrow and steep, with their floors well below sea level. It is generally believed that glaciers created the fjords found in Norway, New Zealand, Chile, and Alaska. But there is much evidence to suggest they are of tectonic origin and not hollowed out by glaciers. These fjords and the Great Rift are intimately connected with the deep-sea trenches and to the amazing crack and fracture zones that blanket our ocean's floors. Most of the fjords and geological features of the Great Rift and the undersea fractures and trenches are young, geologically speaking, and the consensus is that *they happened on a worldwide scale and at the same time.*

This great crack and fracture zones resemble a 20-sided polyhedron. For such a pattern to occur Earth's crust had to expand in a uniform way. Allan and Delair describe it as "earth's crust literally cracking apart like the shell of an overheated egg." How could a planet become a cracked egg? By the presence of thermal energy on an enormous scale built up by overheating molten magma below the planet's crust. Enough thermal energy to thrust mountains to the sky and to buckle and collapse vast land areas, causing some to plunge miles deep into the sea.

Some scholars have remarked that for the crack and fracture zones to have been created in that violent manner, Earth's axis would have to have been on a more vertical plane than it is today. We take Earth's size for granted. You can connect via a computer to anywhere in the world in moments or fly to almost any point in the world within 24 hours. But the size and mass of our planet are immense. Only some powerful *external* force could induce an object of Earth's size to generate enough heat to fracture like an "overheated egg" and, while the planet is orbiting, alter its axis.

COSMIC ASSAULT

At the bottom of our oceans and seas lies red clay. The color is due to the iron compounds that make it up. There could be 10^{16} (10 followed by 16 zeros) tons of red clay resting on the sea bottom. The Pacific Ocean holds the most interesting examples. The clay contains layers of volcanic ash and nickel in high amounts. The ash suggests that numerous volcanic episodes occurred during the clay's formation. Nickel is rarely found in seawater and is hard to find on land. Scientists believe the red clay has an extraterrestrial origin, which came from meteoric dust or large-scale meteor showers. During its world cruise in 1947, the Swedish deep-sea research ship *Albatross* calculated that for the red clay to accumulate on the sea floor in such amounts, 10,000 tons of meteor dust per day would have had to fall since the oceans were first created. They also found that much of the red clay lies on *top* of very young lava deposits.

On top and buried within the red clay are countless numbers of irregularly shaped, radioactive manganese nodules ranging in size from a kernel of corn to three feet in diameter. Their combined weight contributes more than 100 billion tons to our planet. Containing high amounts of nickel, they cover the seabed in places like a gigantic gravel parking lot, particularly in the Pacific Ocean and the Atlantic Ocean off the Carolina coast of the United States. Also mixed in with the red clay are magnetic spherules. A little larger than individual blood cells, they are black and rich in nickel. First discovered in the 1870s and studied again in the 1940s, these tiny spheres are most common in the deepest parts of the oceans.

Though scientists have argued about the age of the red clay, nodules, and spherules, four points need to be made: Their numbers and tonnage are staggering, they have a high nickel content, they are mixed together and found on top of geologically recent lava, and they are worldwide. This is not the scenario for a slow, gradual deposit. They can only be extraterrestrial in origin, and when this mass arrived above Earth, their numbers were more than the stars in the night sky.

Along the North American coast from Maryland to northern Florida are hundreds of thousands of "bays." They are in fact craters. Elliptical and egg-shaped, they are almost invisible from the ground. But from the air they can be seen to stretch from the coast deep into the interior. The Carolinas have the heaviest concentration and have given this phenomenon its name: the Carolina Bays. The smaller "bays" are between 200 and 400 feet long. The larger ones average more than 2,000 feet, with some up to 8,000 feet long. A few reach three to seven miles in length and more than two miles across. In Maryland their orientation is almost east-west; as you move south, the direction becomes primarily northwest-southeast until in northern Florida they orient almost north-south. The land farther inland from the "bays," almost to the Mississippi River, has long been known for its abundance of meteorites.

Near Camden, South Carolina, at a depth of 14 feet, are great numbers of large fallen trees. All give the impression that their demise was swift and violent, as if some large object had broken up in the upper atmosphere into swarms of smaller meteorites and swooped down, from a generally northwest direction, to bombard Earth rotating below. Recall also the heavy concentration of meteoric nodules found off the Carolinas coast. A meteoric origin for the "bays" is accepted, though an age of 12,000 years is argued. Some believe they are even younger.

A similar bombardment seems to have occurred in Alaska near Point Barrow. Ranging from one to nine miles long and up to three miles wide, thousands of oval lakes cover more than 25,000 square miles. The site, which also follows the northwest-southeast orientation, has been dated to about 12,000 years ago. The same story is told in the Canadian Yukon, in northeastern Siberia, the Netherlands, and the Beni region of Bolivia in South America. All these locations lie along a curving trajectory that overwhelms any argument of coincidence and suggests a common cosmic origin.

ICE

We have seen that there is evidence for a geological and biological catastrophe engulfing our world 12,000 years ago. Even though "punctuated equilibrium" is accepted in our distant past of millions of years ago, our recent past remains frozen in a uniform world of ice.

The Ice Age was first advocated in the 1830s. Conceived originally as a catastrophic event, it was usurped by Charles Lyell and his uniform worldview. Lyell used it against the followers of Catastrophism and Biblical literalism. The basic argument against a worldwide Biblical Flood was found with the frozen mammoths then being discovered in the Arctic. Because mammoths had heavy coats, scientists presumed that large portions of the northern hemisphere had to have had a much lower temperature than they do today. From there it was an easy step to imagine the land swallowed in a sea of slow-moving ice, with only the highest summits peaking out from this vast frozen landscape. This was christened the Ice or Pleistocene Age. It was not revealed what caused this age to begin, other than to say some type of climate change. Yet it was accepted within scientific circles, allowing those scientists uncomfortable with a Biblical interpretation to support a more uniform process.

Over the next fifteen years the extent of the hypothesized ice spread over the northern hemisphere. Then the Pleistocene Ice Age became two, not one. A few geologists later, it became three, then five, then seven, and more. No, said others, it was only one, a very long ice age that lasted millions of years, but with numerous advances and retreats. These advances and retreats, which now embraced the Southern Hemisphere too, helped explain the debris fields and bone-caves, and the various phenomena now associated with the Ice Age. They include erratics, moraines, eskers, drumlins, and prairie mounds. This Ice Age became unshakable scientific gospel. Yet it is a web of complex, sometimes contradictory hypotheses that has yet to find an accepted cause or even a consistent definition of its extent.

Erratics are boulders, often immense in size and weight (some are estimated at over 10,000 tons, others are miles long,) that are found over the entire globe. As their name suggests, these boulders lack consistency, regularity, or uniformity in their present locations. They have been found more than 800 miles from their geological source. They are seen in long lines, like a convoy of ships, on mountain crests or clogging valleys in tremendous numbers. They are found alone in isolated locations or half buried in the sides of cliffs. They are found on the Mongolian plains and across the Sahara Desert. You can see them in tropical Brazil and Uruguay. In the United States, Oregon, Montana, New York, and New England have their share. The coast of Northern Ireland has many, as do the near-barren Scottish Highlands. In Ireland the erratics are found on the higher summits and not in the valleys below. In Western Canada they sit more than 4,000 feet above the valley floor. Eastern Labrador, in Canada, has erratics embedded in hillsides, as if some giant hand had rammed them into the earth with great force. Those scientists that follow the Uniform way of thinking say the erratics were brought to their present locations by the slow advance and retreat of the various ice ages. In many places the erratics would have had to defy gravity to reach their current locations, and that the part of Labrador (not to mention Brazil and Uruguay) where erratics are found appears to have never been glaciated.

Many erratics have sharp angular features and are deeply scarred. These rough details conflict with the slow process of transportation by glaciers, where the ice would grind down the sharp features over a long period of time. It suggests instead something much more violent. The scars or striations, according to scientists, were the result of ice scraping across them. But these striations can also be caused by wind and sand, pressurized steam laced with grit, and the tremendous eruptive blasts of volcanic gases. Another possible cause is the continuous pressure applied by debris being carried along by rapid and powerful moving water.

Moraines are drift deposits or landforms composed of accumulated debris, sculpted by glaciers and ice sheets. Massive moraines and drift deposits up to 2,000 feet high are known and are thought to be the remains of those hypothetical Ice Age glaciers. Yet today the Greenland and Antarctica ice sheets and mountain glaciers, many in existence for at least 10,000 years, show little of this effect and *none* of its scale. One would also expect to find the miles of moraine/drift deposits in Canada, Spitsbergen, and the Siberian Arctic Islands to have been pulverized, since these lands are understood to have been *under* a sea of ice for ages. Yet these deposits of sand, gravel, animal bones, and shattered forests, many complete with leaves, flowers, fruits, and nuts and rooted in the ground they grew in, stand as witnesses against the assumption that they were victims of an ice age. And ocean tides, severe storms, cloudbursts, and floods can create moraines and drift deposits as well.

In addition, the ice that today covers Antarctica contradicts the belief that the great ice sheets of the Ice Age carved out valleys and canyons across the world. With a noticeable lack of moraines, the ice appears to be protecting and preserving the Antarctic landscape rather than destroying it.

Ice can move. Science has offered a number of detailed explanations on how this happens, and except for the force of gravity, field observations have failed to prove any of them. Gravity, aided by the ice's weight, pulls it down from the mountaintop. A great mass of ice may move across a level surface for a few miles and then stop. The power, or thrust, that moves the ice comes from its source: elevation. For the Ice Age, as scientists describe it, to have occurred, it would have required highly elevated landscapes from which the ice could flow. We have seen that the high mountain ranges we associate with the Ice Age—the Alps, the Rockies, the Andes, and the Himalayas—were all much lower in elevation 12,000 years ago and incapable of producing the conditions needed for an ice sheet. Without elevation, the millions of mountainless square miles of North America, from the Iowa plains to Northern Quebec's flat tundra, could not have been smothered under a mile of ice.

Scientists counter this nonmovement argument with eskers, drumlins, and prairie mounds. All these phenomena are purported to be evidence of retreating glaciers and ice sheets. Eskers are sand and gravel ridges, sometimes hundreds of miles long, which are believed to be the remains of long rivers that once flowed through tunnels at the bottom of the ice or on glaciers' frozen surfaces. Yet they are found in nature winding like a snake or like a current in a massive body of water flowing over uneven ground and even up and over hills and ridges.

Drumlins are elongated hills, some miles long and more than 200 feet high. They can cover thousands of square miles. In New York one area contains nearly 10,000 drumlins. Their streamlined shapes indicate the direction the ice or water flowed; frequently it appears to have been against gravity. Some drumlins were carved out of softer ground, while others are composed of drift: mostly sand and gravel, but also smashed trees, plants, and animal bones.

Prairie mounds are said to be the remains of stagnate ice from the last days of the Ice Age. This phenomenon assumes that unlike ice today in Greenland and Antarctica, which melts away along the perimeters where it is the thinnest, Ice Age ice was different, melting into isolated blocks while leaving debris and creating the raised prairie mounds. It makes for a vivid picture: gigantic blocks of special ice sitting like chess pieces on the prairie, waiting to melt.

It is interesting to note that eskers, drumlins, and prairie mounds were discovered *after* it was assumed there was an Ice Age. Also, all three are seldom, if ever, found in Greenland or Antarctica and never on a scale produced by the so-called Ice Age 12,000 years ago.

Some scientists, however, have begun to move away from this frozen point-of-view and to study the idea of meltwater bursting in huge amounts from the bottom of ice sheets and glaciers. This raises the question of what would cause such a catastrophe. Since the meltwater floods supposedly came from the interior of the ice and not the top or perimeter, the cause cannot be a sunny day. Did the ground heat up under this particular ice sheet? The melting ice sheets and glaciers in Greenland and Antarctica can be of no help in this matter, since they do not show any evidence of bursts of meltwater.

Even though there are scientists who now support the meltwater hypothesis, they still insist on the presence of ice. Yet research into glacier flow, ice sheet dynamics, and subglacial hydrology is filled with problems and cannot explain the phenomena of erratics, drift, moraines, prairie mounds, and bone-caves as effectively and logically as water.

Cold does not necessarily mean ice sheets. Siberia is the world's coldest region (and has been for the last 10,000 years), yet very few areas have ever been glaciated. Antarctica and Greenland, on the other hand, are both warmer than Siberia. Something more than just cold is necessary to create an ice age.

Explanations abound. Among them are continental drift, wandering of the poles, a sliding Earth's crust, oceanic changes, altitude alterations in landmasses, previous ice ages (a circular argument if there ever was one), changes in climate (always a favorite), Earth's orbital changes, and traveling through cold regions of space. The theories are numerous, complex, and raise even more questions. It is here we invoke Occam's Razor. This rule, first espoused in the 14th century by William of Occam, says, "It is vain to do with more what can be done with less." In other words, given a set of hypotheses, the simplest is probably the best.

Ignatius Donnelly, in 1883, saw it clearly, "What caused the ice? Great rains and snows, they (the glacialists) say, falling on the face of the land. Granted. What is rain in the first instance? Vapor, clouds. Whence are the clouds derived? From the waters of the earth, principally from the oceans. How is the water in the clouds transferred to the clouds from the seas? By evaporation. What is necessary for evaporation? Heat . . . If there is no heat, there is no evaporation; no evaporation, no clouds, no rain; no rain, no ice . . . ice on a stupendous scale . . . must have been preceded by heat on a stupendous scale."

Combining what we have learned about the remarkable and puzzling worldwide distribution of living plants and animals, the remains and locations of their long deceased ancestors, the numerous causes for scarred erratics, moraines, and drift, and their final resting places, the mechanics of ice, and the cause of its creation, it becomes clear that the Ice Age, as we understand it and read about it in books, simply did not exist.

SCALE

Uniformitarianism, the reigning paradigm, boasts the acceptance of most scientists within the last 100 years. Catastrophism, however, has made a remarkable comeback during the past few decades. Referred to as "punctuated equilibrium," and most noticeably applied to the cosmic impact that ended the dinosaur age, this softened version of Catastrophism is still Catastrophism. The two terms, Catastrophism and Uniformitarianism, like two hands attached to the same body, are in truth parts of the same body called science. Allan and Delair insist that Catastrophism is just a question of scale.

Consider doing the laundry. In addition to dirt and grime there are probably millions of microscopic dust mites inhabiting the dirty clothes. Put the clothes in the washer, turn it on, and watch the water begin to rise. Even before the spin cycle starts, the creases and wrinkles, the topography of the clothes, have been altered dramatically. Every living thing, however small, has been either washed away or crushed in vast numbers in the nooks and fissures of the cloth fibers. Yet when we put the clean clothes back on we remain ignorant of the catastrophe that befell the dust mites.

In a park nearby, rain over the past few weeks has been eroding a hillside. Home to countless generations of ants and insects, this hill now collapses in a sudden downpour with much of it washing away. Once numerous living creatures, however small, knew this as their world. Now it is gone, and they are scattered to the four winds. But at our house, the downpour fell into the rain gutters and flowed down into the storm drains. It is just another rainy day. Now consider the dramatic, volcanic eruptions of Mount St. Helens in the United States and Mount Pinatubo in the Philippines. They were both life-altering catastrophes on a local scale. Or consider the hurricanes that one after another violently hammer the southeastern coast of the United States. These local catastrophes are embraced by Uniformitarianism.

Now jump to a point just beyond our Milky Way where you can see our galaxy slowly pinwheeling through space. From this vantage we would notice, scattered around the heart of the galaxy and along its spiral arms, about 40 stars going nova per year. Their explosions would consist of tremendous waves of gases shot off into space. These are distant events and not much to concern us. We would also witness a supernova every 30 years or so. Here, in an incredible explosion more luminous than the light of millions of stars combined, the life of a star ends as its matter is blasted in all directions. This is a cataclysmic milestone by anyone's definition, but on a cosmic scale it is a local event. For astronomers such an occurrence would fit into their cosmic Uniformitarianism despite any local consequences. Should one piece, however, of this exploded star matter come careening through our region of space, the uniform conditions that may have existed from time immemorial in our solar system, and the well-being of every living thing, no matter how small, in the system would come face-to-face with the naked power and cold indifference of the universe.

SECOND MEDITATION

The sun is high over the assembly as we break from our journey. Canopies have been rolled open, and everyone is seated in a pleasant shade. The cranes, now inquisitive about us, perch on rocks that were once, long ago, washed by primordial seas. The short, sharp cries of barking reveal dogs are close by.

As we have seen there is much evidence to attest that something terrible happened on Earth 12,000 years ago. The geology of our planet speaks with authority on the titanic upheavals and collapses of Earth's crust, the abundant evidence of a cosmic visitor, and the misconceptions of the Ice Age. The fact that all the evidence and events took place about 12,000 years ago suggests not an odd regional disaster here or there or a million years buried under ice, but a global calamity of enormous destructive power.

Most scientists, however, prefer to ignore, dismiss, or misrepresent the evidence. They believe that a catastrophe of the scale we are remembering is not compatible with the current Uniform worldview. Yet the geological evidence cannot be denied, nor can their acceptance of such disasters millions and millions of years ago. Recall the dinosaurs were killed off by one such disaster. There are also the numerous articles, many written by scientists themselves, which describe a future filled with potential cataclysms. Who has not heard, read, or seen a movie about a rogue comet or asteroid striking Earth today or tomorrow? Science does not deny that such disasters may happen in the future or have happened long, long ago, but it is selective on when and where catastrophes are permitted to occur. The question is why are scientists selective?

The evidence for an Ice Age is doubtful as well. The hypothesis is based on models that *assume* there *was* an Ice Age. But we have seen that a worldwide flood along with powerful wind could also create the evidence. To remove the Ice Age would be to invite a catastrophic point of view that would again upset the whole Uniform process. So is there another reason for maintaining this wall of ice between the present and our past 12,000 years ago?

Could it be that scientists fear that without the ice, they would have to accept the idea of a flood? And that such a flood would become The Flood? Is science still fighting the battle it began when Uniformity and the Ice Age first arrived on the scene in the 1830s? And has this dread of giving support to Biblical literalists and creationists affected objective, scientific judgment for over 100 years?

We have seen that Uniformitarianism and Catastrophism, parts of the same whole, are reconciled within our reemerging past, and that the Ice Age, as we currently understand it, is no longer necessary. It will, however, return in an important supporting role. It is also clear that our planet, having existed through long ages with little or gradual change, became a participant in a drama that, in a wink of a cosmic eye, left her and all living things upon her changed forever.

6

PHAETON: INTO THE BREACH

At the edge of our solar system lies an interesting anomaly. It is a cloud of Aluminium-26. A radioactive isotope, Aluminium-26 is produced by supernovae. As the offspring of an exploded star, Phaeton was born with all the essential elements for producing the isotope. When he penetrated our system's outermost edge, his electromagnetic and gravitational fields and those of the giant outer planets clashed, triggering a massive energy release. Phaeton was not destroyed in the explosion; instead tremendous energy, fire, and power, accumulated over the long years of its journey, were unleashed. And though this detonation was insignificant compared to Phaeton's birth, the explosion shattered the tenth planet of our solar system into millions of fragments. This explosion created Pluto and Charon, flinging them into their odd orbits, while designing a new ice and stone bracelet to be worn around our sun. Today the bracelet is called the Kuiper Belt. Phaeton captured many pieces of the former planet as satellites, while millions of tons of rock, ice, and gases flowed in long streams behind him. At the same time, newborn asteroids and comets in the thousands were sent streaking like shrapnel toward the planets in the inner solar system.

From Earth the explosion was silent but visible, and ancient observers described Phaeton as having an "awe-inspiring majesty" at this point in its travels. It became twice as bright as the other planets. To the Greeks, Phaeton became the Son of the Sun. Two thousand years ago, Pliny the Elder in his *Natural History* wrote, "It was not really a star so much as what might be called a ball of fire."

On Earth few voices expressed concern. Astronomy is an extremely ancient field of study, and the science behind two lenses made of crystals magnifying distant objects may well have been within the capabilities of our ancient ancestors. They may not have been aware of the destruction of a tenth planet, but what occurred next convinced them, had they not had telescopes, of the existence of a ninth planet.

This fireball, now under watchful eyes, journeyed more than a month in its approach to Neptune. Our third largest planet dwarfed the intruder and, with its powerful gravitational field, stretched the gaseous Phaeton so that it seemed to have two heads. This confirms its molten, malleable nature, and the words that come down to us from the past were that Phaeton had grown four eyes and four ears. "He was not fit for human understanding, hard to look upon," wrote the Babylonians, "when his lips moved fire blazed forth." In this confrontation, Neptune suffered as well. Astronomers note that this deep-blue planet rotates faster than it should relative to other planets its size. Its axis has been pulled over to a 28-degree angle, and it follows a deranged orbit that takes it, for years at a time, beyond Pluto. It has five thin rings, the result of one or more objects breaking up after approaching too closely. One of its moons, the heavily scarred Triton, has a retrograde orbit compared to Neptune's other satellites. This suggests that Phaeton was moving head-on into the solar system. Ovid confirms as much in his *Metamorphoses* when he wrote that Phaeton drove through our family of planets on an "opposing course." Neptune's moons may not even be original moons, but remnants of the now lost tenth planet caught by Neptune as they spun past. This is not a uniform planet family, but one that survived a violent close encounter.

It was here, as Phaeton challenged Neptune, that some of our ancient ancestors understood what they could soon be facing. Only immensely powerful gravitational fields struggling with one another could cause the stretching of astral matter and the creation of Phaeton's second head. If Phaeton had had a weak field, it would have been pulled apart. Seeing this and understanding the implications of how a gravitational tug-of-war would affect Earth and her oceans if this fiery object were to approach close enough, there were those among the watchers who believed that they would be safer on water than on land. When the oceans rose to great heights even the tallest mountains might submerge. Seeking refuge on water offered potential flexibility: flowing with it rather than resisting it and being overwhelmed. The skeptical derided them and ignored their suggestions, but it was at this point that construction began on the first arks.

In the Babylonian epic *Enuma Elish*, Neptune, called Ea, is a sky god who warns Utnapishtim (Noah in the Bible) about a coming deluge. Utnapishtim builds an ark, and with his family, animals and seeds of every kind are brought on board. An earlier tradition also has the hero write down all human history, the mysteries of ritual, and rules of human conduct and buries them in the hopes of recovering them after the approaching disaster.

Humanity watched Phaeton's passage for two months, as the intruder lit up the enormous dark distances between the outer planets. The battle was then joined with Uranus. Often mistaken for a star, this blue-green giant of a planet can be seen dimly by the naked eye. Its combat with Phaeton was a spectacle to the anxious eyes that watched from Earth. Amid the intruder's devouring brightness, Uranus, with a powerful magnetic field, launched tremendous lightning bolts of electromagnetic energy into the invader, cleaving off four large pieces. These newborns were clearly seen beginning to swirl around Phaeton. The Babylonians describe them as four great winds. In Roman myth they were known as Blaze, Dawn, Fire, and Flame, the four horses that pulled Phaeton's chariot as he rode across the heavens. In Christian myth they became the Four Horsemen of the Apocalypse. They joined the chaotic field of debris that circled and flowed behind Phaeton.

Uranus, for its part in this ancient battle, today resembles a wounded soldier lying on his side. Its axis, pulled over by Phaeton, is tilted to a startling 98 degrees. All its moons revolve around the "tilted" equator, resulting in an entire planetary family being perpendicular to the rest of the solar system. Four of the moons are heavily cratered, confirming that a dramatic bombardment took place. Two others exhibit volcanism, suggesting that their internal tidal forces of magma erupted to the surface. Another two moons are small balls of ice and rock, while Uranus's rings are the remains of still other small-sized objects that broke up in its Roche field.

On Earth, work continued on the arks and with the gathering of those things considered important to preserve should disaster strike the planet. People sought safe havens for this knowledge: in a mountain or in labyrinths of caves under the earth. And there were those who thought these actions were excessive and did not believe there was any danger.

The Shining One moved deeper into our solar system. Another 40 days passed before Phaeton entered the realm of Saturn. This was the first time both combatants could be seen easily by everyone on Earth, and at this point the *Enuma Elish* simply states Phaeton "made fear." In a number of traditions, from Italy to India, Saturn was considered to have been the Lord of the Golden Age, but fear led him to devour his children. These children in fact were not Saturn's, but the flotsam Phaeton carried from his previous deadly encounters. As the planet and the invader traded luminous discharges of electromagnetic energy across the darkness, Saturn's powerful gravitational field, like arms sweeping across space, tore objects away from Phaeton's grip. The planet's dense atmosphere swallowed some whole, while others were seen exploding deep within the layers of clouds. The devouring of "children" supports the view that optical instruments were used in observing this drama. (Twelve thousand years later, in 1994, we witnessed through telescopes a similar event when Comet Shoemaker-Levy broke apart and was devoured piece by piece by Jupiter.) Saturn ripped apart numerous other satellites from Phaeton, forming the remarkable bands of rings we see today. At least one, Phoebe, avoided destruction and became a moon that orbits in the opposite direction from all the rest. Phaeton struck back, wrenching the moon Chiron away and sending it into an eccentric orbit between Saturn and Uranus.

Earth-bound observers with telescopes watched Saturn's great cloud belts shift as the planet's axis, like the axle of a giant millstone, slipped, and this lord among worlds, more than 90 times the size of its attacker, was pulled over 26 degrees by Phaeton's awesome power.

For untold ages life both on Earth and in heaven was quiet, uniform, and uneventful. Now this vast celestial battle stirred in many who watched a mounting apprehension and a sense of danger. Did it enter into their dreams? Were others still convinced that Phaeton posed no threat? It will never be known if they attributed this heavenly conflict to the cold, unmerciful working of the universe or to some god or gods. But what happened next caught their breath, like an answer to a prayer. Phaeton stopped.

Rotating furiously in a black ocean, the little sun was held in place for days by the combined gravitational fields of the planets it had just faced in battle. One of Saturn's ancient titles, the "Bringer of Inertia," was surely earned in this encounter. This temporary reprieve held until Neptune, Uranus, and Saturn, scarred and wounded but continuing in the momentum of their orbits, began to move out of range. Straining in the opposite direction, Phaeton felt their lessening grips and the probing fingers of yet another cosmic body. But it was not the pull of Jupiter, already far ahead on another endless circle around the sun.

And here is one of the great "ifs" in history. If Jupiter had met Phaeton face-to-face what would our world be like today? But it was not Jupiter who now reached out with its first tentative and soon unbreakable hold upon the planet killer. It was the Sun. And from the perspective of Earth-bound viewers, Phaeton once again began to move. Passing behind Saturn, Phaeton spawned a fleeting eclipse, like a brief kiss sealing his fate, and then turned toward the Sun. And Earth. No wonder ancient Chinese astronomers called Saturn the "Genie of the Pivot."

Without Jupiter to obstruct his passage in this growing war, Phaeton now crossed into an empty region of space. Nearly a month passed before he reached the next inner planet. By this time the Shining One was bright enough to be seen during the day. At night the first comets began streaking across Earth's skies.

Humanity reacted to the comets with a mixture of wonder, excitement, and a stab of dread. One had only to turn and see Phaeton, dazzling in its brightness, to believe that these streamers of light might be more than just celestial fireworks. They might be harbingers of something terrible. And even if the watchers did not know of the planetary cataclysm at the solar system's rim or what their immediate future held, for the next 12,000 years the sight of a comet would be an omen of calamity.

7

OUR ANCIENT ANCESTORS, PART 1

Standing under the noonday sun, a man blocks the way before us. White-haired, balding on the top, he has bushy eyebrows that shade worried eyes and a robust snow-white beard that cascades down to his chest. He is Charles Darwin, and in 1859 his *Origin of Species* appeared on the scene to complete a triumvirate of new science. Darwin's slow gradual evolutionary process joined Lyell's Uniformity and the popular Ice Age hypothesis to stand as a bastion of a modern worldview and as a bulwark against the implacable Creationist foe. But even as Darwin was writing his history he was worried. He was missing links to complete his masterwork, and he hoped that sometime in "future ages" these links would reveal themselves and confirm his theory.

Entering the 21st century, his grail has yet to be found,

and his hypothesis, like *Titanic*'s iron hull corroded by rust at the bottom of the sea, is becoming a ghost of its former self. Yet his influence remains powerful. Today we suffer from historical Darwinism, a view of history, as dogmatic as any in science, which insists that our past is a series of progressive linear events. From the Stone Age to the Space Age, it claims that our cultural history evolves gradually from the primitive to an ever more advanced civilization.

This view of history has kidnapped a time scale, understood in biological science to span millions of years, and squeezed it, like an accordion, to limit the recorded history of mankind to only a few thousand years. (This is considered "Darwinian" because it evolves from primitive to advanced in a linear fashion not allowing for steps backward.) It accepts that formal science began "suddenly" with the Greeks; mathematics and astronomy arose "suddenly" with the Egyptians and the Chinese; and civilization arrived "suddenly" in Mesopotamia. We look at 5,000 years of recorded history and believe that is as far back as civilization goes. One step further, and we have descended into the Neolithic, the New Stone Age.

Alexander Marshack, a Harvard paleolithic archeologist, in *Roots of Civilization* disagrees. He found "no essential difference . . . between the first fully modern man of some 40,000 years ago and ourselves either in brain size or skeletal measurements." And "as far as the human brain is concerned, the process of preparation, search and research, recognition, comparison, analysis and deducting is the same today as it was 40,000 years ago." Yet science leads us to believe that civilization, as we understand it, developed only in the last 5,000 years, and prior to that, for 35,000 years, we followed after herds of mammoth with clubs and spears and lived in caves.

Marshack believes the "suddenlies" in Greece, Egypt, China, and Mesopotamia could only have come about "at the end of many thousands of years of preparation."

Ancient Egypt is an interesting example. Egyptologists acknowledge that from its beginnings Egyptian civilization already possessed mature art and writing, architecture and sciences. One of Egyptology's founding fathers, Wallis Budge, admitted that much of the Egyptians' skill and knowledge was borrowed from an "exceedingly ancient source." Yet today's Egyptologists choose to ignore this anomaly. To address it now would reveal the flaw in our linear thinking. Maverick Egyptologist John West, in *Serpent in the Sky,* explains the anomaly simply: "Egyptian civilization was not a development, it was a legacy."

In geographical maps we find more signs of advanced knowledge from the past. Maps studied from the 1300s by A. E. Nordenskiöld and then up to the 1500s by American anthropologist and geographer Charles Hapgood reveal a sophistication far beyond the technologies of their times. They both found that many of these maps were based on even older sources, and that the charts displayed a remarkable knowledge of geography. The earliest maps, called Portolans, from the 1300s, were found to be much more accurate than any maps produced during the next 200 years. The Portolans were also superior to the maps of Ptolemy, the ancient world's most famous geographer. His charts, constructed around AD 200, were later lost and rediscovered in the 1500s. The Portolans, along with the High Gothic cathedrals built during the same time, are anomalies whose origins puzzle modern scholarship.

Hapgood's research revealed that many of these maps are drawn on either spherical or plane trigonometric projections, both necessary in constructing complex maps. He also found that the maps are accurate in their representation of longitude. Finding longitude is not easy, and a reliable method to do so was not "discovered" until the 18th century. Many of the ancient maps were also found to have their prime meridians, the site from which other longitudes, east and west, are reckoned, located in Egypt.

All these old projections of Earth imply the existence of an unknown people who, in deep antiquity, understood trigonometric mathematics, had instruments capable of precise measurement, and possessed considerable navigation skills. These remarkable people, however, do not fit into our linear view of history and so they and their maps are ignored or dismissed. Yet they suggest that historical Darwinism is a smoke screen to hide our immense ignorance of humanity's cultural history.

Now a new history is becoming visible through the smoke. Two archeological sites, Catal Huyuk in Turkey and Mehrgarh in Pakistan, are recently discovered ruins of urban cultures, anomalies from the past, insisting on recognition. The first, Catal Huyuk, reveals an unparalleled understanding of technology that confounds us today. We do not know how the people who lived there polished their obsidian mirrors or how they drilled through that hard black stone to create jewelry of the highest caliber. Evidence also indicates that Catal Huyuk was an egalitarian society with a mature religious life. It existed more than 8,000 years ago. Mehrgarh existed at about the same time and was five times Catal Huyuk's size. It was an agricultural center, with many domesticated animals, and had a population of 20,000 people. Egypt, at that time (8,000 years ago) is believed to have had a *total* population of only 30,000 people. No known culture would approach either site's sophistication and urban organization for another 4,000 years. They are anomalies that breach linear time and suggest the possibility of a profound antiquity for our species.

In North America, where historical Darwinism is held high, the orthodox view holds that migrants from Asia crossed the Beringia land bridge no earlier than 12,000 years ago. This is a remarkable feat because much of North America, according to the same orthodox view, was supposedly under a vast sea of ice. Since it was necessary to get these people through, to conform to the paradigm, an ice-free corridor thousands of miles long was suggested. Passing through two parted ice sheets, like Moses and the Hebrews crossing the parted Red Sea, these first Americans were able to wander over the entire New World. In only 500 years they managed to travel more than 11,000 miles, from the Arctic to Tierra del Fuego, across vast chains of mountains (not to mention ice sheets), bone-dry deserts, and dense rainforests.

Science prefers to support this view and ignore the discoveries of Thomas Lee, who in Canada during the 1950s found that people had been in the new world for more than 30,000 years. On Manitoulin Island, in Lake Huron, Lee, who worked for the National Museum of Canada, uncovered dozens of stone implements. The level at which they were excavated indicated a date older than the orthodox view of 11,000 years. Geological evidence supported Lee, but the scientific establishment did not. Because he challenged the prevailing belief, Lee was blacklisted, his work was dismissed, his papers suppressed or lost, and he was forced, in the end, to resign his position.

In the winter of 2000, an early human site at Cactus Hill, Virginia, was dated as 17,000 years old. And in South America at Monte Verde, Chile, multiple sites with human occupation have been dated at more than 12,000 years old. Our current worldview resembles not so much a bowl filled with our scientific knowledge as a colander with anomalies streaming through.

With the linear foundations of science disintegrating, we begin to hear the voices of our ancestors. Contrary to what we are taught in school, man did not necessarily reach the new world via the Beringia land bridge. The Aztecs, long before they built their magnificent city of Tenochtitlán in old Mexico, came from a distant island homeland they called Aztlan. The Hopis, now living in the American Southwest, originally came from Muia, an island in the Pacific. Many tribes in the Pacific Northwest, including the Haida of the windswept Queen Charlotte Islands of British Columbia, also crossed the Pacific by boat and canoe. This consistent memory of a homeland in the Pacific is also common through the Central American rainforests and down South America's western coast. On the other side of the continent, the Maya, in Yucatan, believe their ancient fathers arrived from over the eastern sea.

No matter from what point of the compass they came, their memories fail to mention passing through walls of ice. They arrived by sea. In boats. Their traditional histories are also in agreement that this arrival occurred after a worldwide catastrophe. And they all recall that before this terrible event, they lived in a world so utterly different that when we look back 12,000 years we seem, through their stories, to catch a glimpse of paradise.

MYTH

Myths are our ancestral memories. Transmitted through generations, they share their ancient knowledge of how we once viewed the world. Science does not accept these myths, traditions, legends, and literatures as historically accurate because they can seldom be verified, duplicated, or counted. Instead myths are viewed symbolically. Through the symbols, metaphors, and allegory of myths, stories are told that are considered both entertaining and instructional but not historical. Mythologist Joseph Campbell is the most well known scholar of symbolic interpretation.

Historical interpretation, a less common method for understanding myth, was established around 300 BC by the Greek mythographer Euhemerus. In this view, mythological characters and events are based on historical people and occurrences. Homer's Troy was believed to be a myth until Heinrich Schliemann uncovered it in 1871. The wealth of the Egyptian pharaohs was believed to be a myth until Howard Carter discovered King Tut's tomb in the 1920s. Much precedence exists for the truth of myth, but it is a difficult task to find the kernels of truth in memories that may have accumulated a lot of dust over the centuries or millennia. Immanuel Velikovsky, whose views have been widely criticized by the mainstream, is among the proponents of historical interpretation.

Science prefers the symbolic method and ignores the historical implications, that is, until they are proven. It could be that the symbolic and historical views of myth are both correct. Just like any great literature, from the Bible to Shakespeare to James Joyce, myths have multiple perspectives. They are not just words telling a single story. They are more like holograms of our past. Their multidimensional nature appears when our deductive faculties are focused.

Otherwise we see a distorted picture revealing, like the broken stick in the pond, how we can deceive ourselves.

We have long understood that mythology and our spiritual comprehension of the universe are related. Through research in physics, science is coming to realize that it is also a part of that story. Joseph Campbell said that the world's myths "resemble each other as dialects of a single language." In order to fully comprehend history it may be that the mythological, spiritual, and scientific perspectives—the dialects of interpretation—must harmonize, not contradict.

PARADISE

Sometime during our lives we have all imagined Paradise. "Once upon a time," the stories begin. One of the most compelling types of myth we have is the consistent worldwide memory that humanity once lived in Paradise. Some say that we should seek Paradise within, since it never existed outside us. Is this paradise only wishful thinking from a long-suffering species, a trickster apparition that teases us with what we never had, once upon a time? Or could there be historical truth buried deep within the myths?

We have discussed the fact that our past, 12,000 years ago, was a dynamic time. The world was *not* encased in ice. Instead it was a climatically temperate world filled with abundant plant life. Vast herds of animals and expansive forests covered the continents. Myths from every corner of the world support the biological evidence. Most familiar to us is the Garden of Eden, with its fruit-bearing trees and sweet water. In Greece's Golden Age the "untilled earth bore its fruit, and the unploughed field grew hoary with heavy ears of wheat." Spring was eternal here as it was in the Egyptian First Time, known as the Tep Zepi, when everything good was in abundance.

Throughout Polynesia, across to the high Andes, up to the Great Plains of North America, and from the deep rainforests of Africa to the steppes of Asia, the memories are the same: We once lived in a world of abundance and peace.

Other myths of Paradise describe animals being less fearful and aggressive. Today, one can journey to Ecuador's Galápagos Islands and look into a sea lion's eyes, mere inches away, see absolutely no fear and realize that a different world can exist and has existed before. It is an emotional discovery going beyond laboratory science, and our scientific and religious prejudices will resist, but our spirits connect.

In these myths, many skills and qualities are attributed to our ancestors who lived during this age. They spoke a single world language, had the ability to communicate with animals, and were, for the most part, vegetarians living in harmony with all things. In addition, they were luminous, long-lived, and wise.

How much of this is believable? Could all these stories be fictional, a symbolic desire for something we do not have? Skepticism is important; to accept it all without reservation would be as myopic as to reject it outright. Yet these myths relay consistent information and come from hundreds of cultures across the planet. Some have, no doubt, influenced others, but there is too much similarity to account for all the myths in a symbolic way.

LANGUAGE

Today there are, depending on the criteria, up to 10,000 languages spoken on Earth. However, many of our most ancient myths tell of a single world language. The fact that so many diverse traditions throughout history share this belief, from the Hebrews of the ancient Middle East (prior to the Tower of Babel) to the Chins of Southeast Asia and the Maya of the Americas, suggests the myths may be true. Linguist Joseph Greenberg narrowed the number of languages, which share a common heritage, down to 17 major groups. Then in 1995, Merrit Ruhlen and John Bengston proposed that a common language, Proto-Global or Proto World, spanned the entire planet more than 12,000 years ago. For example, the word for *water* in Indo-European (our linguist group) is *aqua*. It is found in Afro-Asiatic groups as *ak(k)a*. In the Altaic family, which includes the Turkish languages, Mongolian, and Japanese, it is *aka*, and in the Amerindian family it is called *waka*. Simple as they may seem, these similarities are found over the entire globe. They cannot be explained away by borrowing from other cultures or by coincidence.

Another skill myths across the globe mention is the ability to communicate with animals. Many African myths remember that we once understood many animal languages. The Greek Aesop believed that during the Golden Age animals understood the use of words. This belief is reflected in his collection of fables, gathered from many older sources.

Today this "skill" is still very popular and can be found in the numerous talking animals in our literature and film. Also, many people today speak of an ability to understand and be understood by their pets. While some people shake their heads, others know a connection is being made. The successful research of the last few decades initiating communication with dolphins, apes, chimpanzees, and most recently with parrots supports the truth of these age-old myths.

Language was a turning point in our evolution, like the winter solstice that brings light to a darkened world. It allowed us to learn and share what the universe, so marvelous and mysterious, had to teach us. It was the glue that united us in an ancient global village. Would *that* be worth remembering?

HEALTH

As we continue this line of thinking, we find many myths that speak of people being vegetarians. Many myths that describe communication between species mention a close harmonious relationship with animals as well. In the Garden of Eden, Adam and Eve are given all the seed-bearing plants and fruits to eat. Not meat. It is only after the Flood that mankind begins to eat animal flesh.

We have been meat-eaters, according to science, from time immemorial and certainly within the last 10,000 years. Vegetarianism has only made global inroads in Western culture within the past 30 to 40 years. Why would these old myths recall that we did *not* eat meat? It is unlikely that such culturally and geographically diverse people would all come up with the idea of vegetarianism and then add it to their myths. Though it is widespread in our myths, it is not universal. The Aborigines of Australia recall that game was abundant in their Golden Age, the Great Dreamtime.

A recent field of scientific study deals with the blood that our ancient ancestors, generation after generation, passed down to us. It uses blood types to show how those ancestors spread across the planet and adapted to challenges they faced. Type O is the most ancient. It appeared between 40,000 to 50,000 years ago. Recent research into the history of our blood types has produced some interesting ideas. One of these is that people with type O appear to thrive on meat. Type A seems to have emerged sometime between 15,000 to 20,000 years ago, and those people may live a healthier life if they eat no meat. This perhaps reveals the ancient change to a more vegetarian lifestyle. Type B is the tantalizing one however. It appeared mysteriously on the world scene 10,000 to 15,000 years ago, and people who have this blood type are believed to have a healthier life if they balance meat and vegetables in their diet. The global disaster we are remembering 12,000 years ago falls within these dates and may reflect a switch back to meat after much of the plant life was wiped out.

Recall that our ancestors inhabited a world of vast forests and vegetation with many animal and plant species that are now extinct. It can be assumed that the edible plants and herbs, grown in soil fertile beyond our imagination, were much higher in nutritional value than the nutrient-depleted meat and produce we consume today. Also, prior to the disaster, our atmosphere contained up to 15 times the current concentration of carbon dioxide, which stimulates circulation and oxygenation in animal brains. With improved circulation, oxygenation, and better nutrition our ancient ancestors' brains may have functioned better than our own.

The additional carbon dioxide also promotes plant growth. We know from their remains that plants *and* animals 12,000 years ago were much larger than they are today. Many traditions recall a time when giants (humans sometimes included) walked the earth.

One of the most consistent benefits of good health found in Paradise myths is long life, sometimes described as immortality. The Bible records that the patriarchs, the fathers of the human race, lived for thousands of years, while other spiritual traditions view our present life as one stop in a greater cycle of many lives through reincarnation. Such memories and beliefs led to stories of immortality in mythical memory. Are they fictional stories, or can there be some truth in them?

Another benefit recalled in worldwide myth, and perhaps explained in part by good health, is that we were luminous. Adam and Eve, in esoteric Christian traditions, noticed their loss of body light after they were escorted out of the Garden. Many Asian myths, among them those of the Tibetans and the Siberian Kalmucks, recall how their ancestors in Paradise radiated light. Today we read about historic individuals who were luminous: Moses, Jesus, Christian saints, and Eskimo shamans. The Christian mystic Meister Eckhart called it the "divine spark" possessed by all human beings.

Up to this point there has been little scientific, verifiable proof to support the existence of these marvelous human attributes. They are, after all, just myths. Yet we find them consistent within each individual myth and within the worldwide body of Paradise myths. Their very existence, in a proven physical paradise, adds support to the biological and geological evidence. And considering what follows after the disaster, such amazing skills and traits *would* be worth remembering. However, such conditions would not be achieved "suddenly," but learned through experience over the course of thousands of years in an environment easily conducive to human life. The question is, does any evidence exist to support this ancient knowledge?

SPIRIT

Woven within our lives, and stretching back like long threads across time, are five beliefs that have come down to us from an unknown fountainhead. In *Lake of Memory Rising,* author and humanist William Fix noticed that these beliefs appeared first on one continent and then another, during this century or that millennium. Through spiritual traditions, religious movements, and moments of enlightenment, men and women like us have shared these beliefs as revelation. Today, they are embedded in all of our major religions: Buddhism, Christianity, Hinduism, Islam, and Judaism, reminding us of our potential to exist in a more spiritual world.

The first is a near universal belief in a creative source. Arthur C. Clarke once wrote that there were 9 billion names for god, titles for our unfathomable beginning, or some divine being that set the universe in motion.

Second, we find, extending from ancient Egypt across 2,000 years to Jesus of Nazareth, and then across another 2,000 years to the current Amerindian cultures of North America, an acknowledgment that humans are all from the same family. We are brothers and sisters. There are no exceptions, no exclusions.

Third, we have all existed, throughout time, in the same spiritual universe—that is, a universe where we share many of the same rituals and symbols in our myths, daily beliefs, visions, and dreams. For example, rituals of birth, rites of passage, marriage, and death have common elements throughout history and cultures. Others may not appear the same at first, but they serve the same purpose. Numerous and diverse spiritual paths can be followed, but in the end these paths will join together and lead to the same place: liberation, spiritual enlightenment, nirvana.

Fourth is the One Law. It is found in almost every religion in the world. "Love your neighbors as yourself," said Leviticus. Confucius says, "What you do not want done to yourself, do not do to others." This law has come down to us as the Golden Rule: Do unto others, as you would have them do unto you. It is the law of reciprocity. It is karma. It is wisdom pure, simple, and straight from the heart. In a world where this is held to be the highest truth and practiced as such, we would find no greed, hate, racism, or violence toward any living thing.

The fifth is revelation. Our current concept of truth's disclosure has come to be understood, in the last 2,000 years, as the Logos, the divine Word of God. But *this* interpretation originated almost 10,000 years *after* the event we are remembering. Our ancient ancestors, 12,000 years ago, realized revelation not as the Word, but as an effect of their everyday action of following the One Law, doing unto others, as you would have them do undo you. "The spirit," William Fix writes, "will not cease revealing, nor wait upon a time; whenever it finds channels, it will pour forth." From this comes the final insightful discovery: A connection to that spirit is always within us.

Did all these same five beliefs rise independently in remote and diverse locations, times, and cultures? Or did they flow from a common source, the Golden Age of our myths?

DOMESTICATION

Within the last 2,000 years the only animal to be domesticated has been the rabbit. French monks accomplished this during the sixth through the tenth centuries. Domestication can be defined as the genetic manipulation of a wild plant or animal for the advantage of humans. So who domesticated all those other animals besides the rabbit?

Animal domestication is believed to have occurred within the last 10,000 to 16,000 years. Most of the dates hover around the 10,000-year mark, a time that coincides with the end of the illusionary Ice Age. You can read any number of books on this time period and its populations of cavemen, mammoths, saber-tooth cats and so on, and you will find almost nothing about the sheep, goats, pigs, and chickens that inhabited the world with them. But after the ice melts, "suddenly" out of nowhere, these animals appear already domesticated and part of the web of human life.

We know, from the incredible number of bones found throughout the world, that the wild herds in the Late Pleistocene, 12,000 years ago, would dwarf today's great African herds and the sea of bison that existed in North America 200 years ago. It is possible that within those teeming ancient gatherings of mammoth, wooly rhinoceros, camel, and horse, moving across a bountiful landscape would be reindeer, cattle, sheep, and goats.

Over time our ancestors would have looked at their Earth and seen seeds, which produced the grain they ate, scatter in all directions in a sudden wind, or the animals that supplied them with material for their clothes, milk and perhaps meat for their bellies, wander off to the far side of the hill.

Then one day someone decided to make the odds better. Perhaps it was a desire to cease wandering or simple curiosity that urged them to confront that chasm between species. Our ancient ancestors had the intelligence and the time necessary to domesticate those first wild animals. Such a process, over generations, had profound effects on humanity's development. Our very concept of ourselves and our relationship with other living things changed radically.

The use of animals for transportation, food, clothing, and companionship necessitated new systems of classification, celebrations, taboos, and traditions. Our language increased with each animal taken into the fold. Veterinary medicine was born. Instead of chasing after the animals when we needed something, they would be with us from now on, and they offered us a dividend: From their manure came a rich source of fertilizer for our expanding garden.

It is a mystery, as well, when the first plants were domesticated, probably sometime in the last 10,000 to 15,000 years. Surviving evidence indicates cultivation took place in highland regions rather than along rivers and valleys. The reason for this will become evident soon.

The wheat, rice, corn, and barley so necessary to our survival today were designed with patience and perseverance. They were created so their size increased and their taste improved, from bitter to sweet. Their seeds would no longer disperse in the wind, and they could be harvested at regular intervals.

Human hands also transformed nuts, apples, and olives. Citrus fruit were persuaded to loose their thorny protective covering. One, the marvelous banana, is sterile and cannot reproduce without human assistance. The names of those ancient genetic engineers are lost in time, but they did leave us a tantalizing legend. One variety of bananas is known as *Musa paradisiacal,* the fruit of paradise. It is said to have flourished in the Garden of Eden.

Supporting the belief that these achievements have an enormous antiquity are a large number of worldwide Flood myths describing people with their animals seeking safety on mountaintops, in caves, and on arks. As we will see shortly, these people, particularly those seeking refuge in arks, were also advised to include seeds of their most precious plants. Though horticulture was probably not widespread during the Golden Age, certainly having domesticated plants and animals would make the terrible labor that would soon follow—rebuilding a ravaged, broken world—easier.

MEDICINE

Across uncounted years, through accident, observation, and trial and error, our ancient ancestors found that beyond spicing up a head of cabbage, plants had medicinal value.

All the thousands of botanical remedies that we employ today (either in natural forms or chemical reproductions) were discovered once or many times in our history. One can only guess at the cost, in both human life and time, involved in these ancient discoveries. The joy of a successful cure and the agony of failure are experiences everyone, across time, can understand. But in the end, our ancient ancestors' medical knowledge has served humanity well for more than 12,000 years.

When a patient's condition required drastic methods, our ancient ancestors were up to the challenge. Trephining or trepanning is a surgical procedure that involves cutting a hole, up to two inches in diameter, in a skull. It was used to remove splinters or pieces of bone lodged in the brain, to relieve pressure, or some theorize, to treat mental illness, headache, or epilepsy. It was accomplished with flint scrapers or obsidian stones so razor sharp they can still be used today. A plug of animal bone was sometimes used to seal the incision. Such an operation requires sanitary conditions, as it exposes sensitive brain tissue. It may also cause considerable loss of blood. We know many people survived this delicate operation because their excavated skulls show healed bone. This surgical procedure has been found in skulls more than 12,000 years old and is acknowledged to be worldwide in extent. Evidence of this operation is found in Europe and across Asia and the Pacific to South America. It is today still practiced successfully in parts of Africa.

In 1969, Russian researchers in Central Asia uncovered 30 human skeletons, all at least 12,000 years old. A number showed evidence of successful skull surgery, while several showed features that indicated what surgeons call a "cardiac window," an opening in the chest that is necessary for open-heart surgery. Why were such operations performed? The releasing of evil spirits is popular among those who choose to see these skeletons as examples of a primitive cave-people. Others, however, acknowledge that such complicated surgery demands an intimate understanding of human anatomy, medical procedures, and genuine technical skills. Again, evidence indicates that the patients survived the operations.

From Egypt, the Ebers Papyrus, believed to be 5,000 years old (from when Egypt "suddenly" appeared fully developed on the banks of the Nile) and over 100 pages long, is a medical text offering diagnosis and treatment for a variety of skin diseases; dental, dermatological, and gynecological conditions; and ear, nose and throat diseases. The shorter Edwin Smith papyrus displays a highly complex knowledge of diagnosis and remedies for treating wounds, fractures, dislocations, and bruises. When this document was first deciphered it was smugly called "sewage pharmacology" because its medical treatments included the use of droppings and urine of pelicans and hippopotami among others. In Asia, the ancient Chinese used bat excrement as a cure for night blindness, a reduced ability to see at night or in dim light. A chemical breakdown of the bat droppings reveals high quantities of vitamin A, a deficiency of which causes night blindness. Only in the last 100 years has our science realized that these "sewage" agents are rich in natural antibiotics.

In India, various battle wounds have been cured and artificial metal limbs have been created for more than 4,000 years. We have had such skills in the West for less than 300 years. This rich wellspring of Indian medical knowledge comes from the *Vedas,* one of the most ancient spiritual literatures in the world. It has come down to us as *Ayurveda,* the life science. In addition to its spiritual side it contains practical information for healing wounds and also a vast library of naturopathic medicine. Yet Western medicine has long viewed ancient medical knowledge from other cultures as superstition.

China's medical system is also of great antiquity. A legend tells that the Yellow Emperor, Huang Ti, wrote an authoritative work on internal medicine nearly 5,000 years ago. Acupuncture, the most famous body of Chinese medical knowledge, goes back more than 4,000 years. It is a style of medicine based on vital energy currents that move through the human body. Thought to be unique to China, it is slowly gaining acceptance in the West despite Western science's initial conclusion that it was a magical process to drive out evil demons and our current inability to scientifically explain how it works.

Acupuncture's origin has recently come into doubt. In 1991 a mummified man was found at the foot of a retreating glacier high in the Italian-Austrian Alps. He became a world celebrity and was christened the Ice Man, or Otzi. He lived and died more than 5,000 years ago and has provided science with a unique window into the past. One interesting discovery was 15 groups of small bluish-black tattoos located on Otzi's back, right knee, and left ankle. These markings match the locations of traditional acupuncture points. Other analysis revealed that Otzi suffered from arthritis in his spine, knees, and ankles, and that he had worms in his digestive system when he died. His tattoos coincide with the acupuncture locations used for the treatment of backache and upset stomach.

Did Otzi travel to China and receive treatment, or did the Chinese come west to teach this system of medicine? Such a sophisticated method could not have "suddenly" appeared nor would it have appeared in numerous places independently. Rather, acupuncture, together with the 12,000-year-old procedures for trephining and open-heart surgery, represent a growing body of evidence indicating the possibility of a far older heritage of knowledge for our species—a species that had the time, intelligence, and skill to develop such life-saving medical procedures and then, after the disaster, sought to pass that information down to future generations.

8

PHAETON, THE PLANET KILLER

Traditions from Africa, Asia, the Americas, and across the Pacific recall that an awful drought preceded Phaeton's arrival. Earth, disturbed by the approaching magnetically charged intruder, was beginning to react: Her internal magma tides continued to rotate normally, but the movement of her surface crust slowed slightly. This created friction and increased the thermal energy rising from deep within the planet. The increased heat evaporated moisture and turned the vast green forests and plains brown.

Phaeton's course now took it to meet the fifth planet. What we today call the Asteroid Belt existed 12,000 years ago as another planet in the House of the Sun. The *Enuma Elish* called this planet Tiamat and her single moon Kingu. Astronomers, after determining the number of asteroids within the belt, suggest that this missing planet had 90 times the mass of Earth; it was a little smaller than Saturn. As Phaeton drew closer, its gravitational power pulled Kingu away from its parent. When Phaeton and Tiamat finally clashed, amid blinding discharges of electromagnetic energy, a new element was injected into this fierce confrontation: sound. Before this distances were too vast for anything to be heard on Earth, but now the silence was shattered with deep booming and acoustical roars reaching across space.

Looking up from their work on the arks, made of both wood and reeds, the startled builders tried to steady their shaking hands. Others gathered supplies to be stockpiled in caves high in the mountains. Earth's vast animal herds moved restlessly among the tall drying grasses and forests. A sense of urgency grew.

Electrical discharges lit up space like a sparkling net that both Tiamat and Phaeton entered. Tiamat's gravitational power caused three more balls of fire and gas to erupt from Phaeton and begin to whirl, like dervishes, around their master. Spitting fire, Phaeton approached. Composed more of solid rock than of gases, Tiamat groaned as huge pieces of her crust were ripped from her surface. The *Enuma Elish* says these pieces, Tiamat's "eleven mighty helpers," were wrapped in halos of light. Kingu, the stolen moon, now reentered the battle, and all three bodies unleashed thunderous lighting bolts at one another. The heavens were at war.

From Earth, amid rising temperatures, drought, and growing clouds, even doubters stared into space and shuddered, as this unbelievable spectacle unfolded. Panic appeared in people's eyes, and not a few wondered if they were living at the end of time.

For Tiamat the end was near. Phaeton's grip slowed her rotation. While her inner core still spun at its normal speed, Tiamat's outer surface slowed to a stop, and the doomed planet began to break up. She roared as her outer crust split open. Phaeton fired blasts into the breach. One of the flaming whirlwinds slammed into Tiamat. The entire crust of the planet rose and collapsed like a bellows. The Babylonians recorded how Phaeton "shot off an arrow, and it tore her belly; it cut through her insides, it split open her heart." He struck again and again, dismembering Tiamat. Whatever atmosphere she had, whatever liquids, gases, and ice, were vaporized in the hellish final minutes or blasted into space and instantly frozen into waves of sparkling crystals. A portion of her crust was shattered into cosmic dust. What was left of the fifth planet was ensnared in Phaeton's sweeping gravitational net and flowed for a time behind the invader like a great tail. Tiamat's "mighty helpers" and Kingu became satellites to the unstoppable planet killer while others formed a hammered bracelet of asteroids and comets around the Sun.

From Earth, Tiamat could no longer be seen in the night skies. She had disappeared. In ancient Greece, Electra, sometimes described as a planet and one of Atlas's seven daughters, left the heavens because she was unable to watch the bloodshed of battle. There are two accounts. In one she is never seen again. In the second she shows herself only occasionally, to mortals, in the guise of a comet.

Tiamat's disappearance produced shock waves that broke over humanity. Before our eyes an unwelcome guest, second only to the Sun in brightness and power, had disemboweled a planet, one that had been a comforting companion across uncounted ages. Crops, orchards, and forests continued to wither, rivers and lakes to evaporate. Great flocks of birds flew about aimlessly, their navigation disrupted, as Earth's magnetic field reacted to Phaeton. Animal herds moved over the land in search of food and water. The sky grew crowded with heavy clouds creating vivid sunrises and sunsets. Nights were without stars. A sense of dread and bewilderment spread over the people.

The juggernaut Phaeton and his largest satellite Kingu were now affecting and distorting the atmosphere of Earth and Mars. Both planets' immense seas of gases, already under stress from growing internal temperatures and electromagnetic activity, were becoming thinner as they were pulled outward, conelike, thousands of miles into space. Crackling with static electricity, winds on the two worlds blew hot and dry. The evaporation of moisture became acute.

On Earth, mastodon, camel, zebra, and other herds, in great shifting, agitated masses, pressed forward to drink in muddy streams, shrinking rivers and lakes. Mirages, not seen before, contributed to disorientation and loss of life. Rumors of rain, the end of the world, hidden sanctuaries, and attempts to take control of arks spread quickly and clashed with pleas to stay calm. Emotions and fears unfamiliar to all were forcing themselves into their hearts and minds.

The little sun and its followers crossed the distance between the ruins of Tiamat and Mars in ten days. Today we know Mars as the red planet, the god of war. But this has not always been the case. In very ancient times Mars was a deity of the fields and marshlands, of spring, fertility, and cultivation. Recent evidence also leads us to a different image of Mars, with data from orbiting Mars surveyors suggesting that it once was a planet rich in water; water that had existed for millions of years and then disappeared. It is reasonable to infer that plant life existed and was verdant and abundant enough to visually impress beings on a sister planet millions of miles away. Only later was Mars transformed into the god of blood and war.

From Earth the heavens, increasingly obscured by clouds, could not hide the fact that as fiery Phaeton closed in on Mars, that planet became brighter. Increased heat had desiccated the rich vegetation and the vast lakes that dotted the planet. As though preparing for the coming battle, the Martian atmosphere swirled with tremendous electrical storms, which sparked fires on the surface that quickly ignited the now tinder-dry landscape.

Due to the pull of Phaeton's gravitational grip, the Martian waters rose to great heights, and the rotation of the planet slowed. Winds whipped into ferocious storms and turned the growing fires into a planet-wide holocaust. The light from this conflagration reflected through the atmosphere and created an enhanced glow to humanity observing from across space. Mars was on fire.

Like two angry bulls in a field of stars, the combatants approached one another. Phaeton crossed behind Mars, creating a fiery halo in a short-lived eclipse. Mars was now on the left as the little sun turned and launched a mighty electrical bolt toward our sister planet. Mars bellowed and fired hundreds of lightning rockets into Phaeton. At the same time the Martian atmosphere faced the onslaught of asteroids and comets from the remains of Tiamat and the tenth planet. Some exploded as they entered Mars's Roche field. Some ricocheted off the atmosphere back into space, while others lacerated the dark Martian skies with flaming streaks and then disintegrated in the chaos. Still others hammered the surface, riddling it with sharp, deep craters. Mars heaved more thunderbolts into Phaeton and Kingu, which staggered from the assault. The last remaining large piece of Tiamat, dragged across space by Phaeton, careened toward the planet. It smashed into the Martian surface, blasting a crater thousands of feet deep and a thousand miles wide. Moments later on the planet's opposite side a towering bulge swelled 6½ miles into the sky and spread itself over 3,000 miles. From this monstrous growth, crevasses, resembling arthritic fingers, quickly opened into chasms stretching for hundreds of miles in all directions. Lava surged through the planet's hemorrhaging crust. Rain crashed down from the sky and floods rolled over the surface, sometimes colliding with lava flows, creating eruptions of steam that rose miles high. The cosmic battle was heard on Earth, but little was seen of Mars as our atmosphere, already cloudy, became agitated with electrical storms that ignited the first fires.

Mars was now lurching in orbit rather than spinning, its axis losing stability. The positive and negative magnetic fields at the poles flipped and then shifted, and for a brief moment Mars ceased rotating and shorted out, its gravitational and electrical fields nullified by Phaeton. In that small moment of time, before Mars began to rotate again, part of its atmosphere, surface crust, and waters were sucked into space and caught by Phaeton's gravitational field. Then with cold indifference, Phaeton moved past Mars and continued on its journey toward Earth, leaving two small, scarred stones that to this day trace wild orbits as moons around a dead planet.

An awful moan, originating from the heavens and sounding like the lamentations of an utterly broken people, pealed over Earth's skies, announced the battle's outcome.

Looking up, humanity saw Phaeton, awesome and unstoppable, its glow burning through the layers of clouds.

Mars was there as well, but no longer would he represent spring and fertility to the people of Earth. When studied hundreds, then thousands of years later, Mars would be smeared in dust and the color of blood. He had been transformed into a red god of war for a new kind of man. Yet in an old Homeric hymn Mars was remembered as a defender of humanity. And at one time, 12,000 years ago, he was.

9

OUR ANCIENT ANCESTORS, PART 2

Evening approaches as we continue along the path. We have left behind the notion that our ancient ancestors 12,000 years ago were just cave dwellers. Much of what we take for granted today—our domesticated plants and animals, our medicines and language, and the spirit of that golden age—are a legacy from those ancestors recorded in myth. Those myths are the many colors and images embroidered in a great cloth that cloaks our mysterious past. The coat shimmers in the hues of paradise and heaven, and reveals the images to be great trees, dragons, the stars above, and more.

ENERGY

The ancient Chinese believed that the flowing energy, which forms the basis of acupuncture, was not limited to the human body. Earth has this energy too. Known as paths of the dragon, or *lung-mei,* these energy lines crisscross the entire planet. In China these paths, sometimes cut into the earth, still connect ancient sacred sites, astronomical observation locations, and high mountains. For centuries only members of the Imperial family could be buried in tombs located on a dragon path, so important its power to control a family's future. This ancient energy system has come down to us today in distorted form as *feng shui,* literally translated as "wind" and "water," the science of alignment where structures and landscapes take advantage of, and seek harmony with, Earth's energy.

Dragon paths have been studied for many years in the British Isles. An Englishman, Alfred Watkins, in a moment of insight in 1921, recognized something that had become lost to memory. He saw, stretching the length and breath of the islands, a web of "old straight tracks" running across valleys and mountains. Sometimes only as wide as the back of your hand or broad enough for a road, they ran straight as an arrow and never around an obstacle as paths and roads do. They linked and intersected crossroads, sacred trees, holy wells, earth mounds, geological features, churches built upon pre-Christian sites, megalithic structures, and circles of stones. These "tracks" of a forgotten world are known as ley lines.

Today we remain ignorant of the purpose of these energy paths and lines, but there are alluring hints. We know that over our world flows magnetic energy in invisible waves. Earth is, in fact, a giant magnet. The energy ebbs and flows in small and great spirals. Like currents of air it circulates all around us. Emanating from the northern and southern magnetic poles, this magnetic energy was acknowledged by a great many ancient traditions.

Plants and animals respond to these vibratory waves of energy. Tomatoes and mushrooms as well as bacteria, crabs, honeybees, and mice, among many other living things, are sensitive to this magnetic influence. The great migratory cycles of animal herds in Africa, salmon finding their way back to the very stream of their birth, swallows that after traveling 15,000 miles return to the same location year after year, all respond to a shared dynamic relationship between themselves and this magnetic energy that washes over our world.

We, being creatures of Earth, are also sensitive to the magnetic energy, though traditional science separates us from our environment and generally chooses to ignore these findings. The religions of Abraham (Christianity, Islam, and Judaism) also share this separatist point of view. But there is a growing body of scientific work that shows Earth's magnetism affects numerous processes in the human body, including the brain and the pineal gland. And thanks to theoretical physics, science is beginning to grasp the sublime complexity of a holistic universe. Scientists are finding a profound interrelationship that reveals we are not apart, but rather a part of this energy that permeates all creation. Yet our ancient ancestors gazing out over fields of wild barley and distant herds of mastodon, giant bison, and camel understood this holistic concept 12,000 years ago.

These dragon paths were no surprise to the Romans when they arrived in Britain more than 2,000 years ago. They had seen them in the Fertile Crescent, North Africa, and throughout Europe. Ever pragmatic, they built their amazing road system on top of many of the old straight tracks. Before the Romans, the Druids inherited the paths, but the full potential and history of this Earth science was a mystery to them.

Across the Atlantic Ocean, the Americas have their dragon paths. In the Four Corners region of the United States—Arizona, Colorado, New Mexico, and Utah—there are numerous paths, sometimes referred to as "interstate highways" by American archeologists, which cut across the arid landscape. Speculation, as to the purpose of these paths, ranges from roads connecting ceremonial sites to local projects established to create social cohesion and unity among local populations.

In South America, in the Bolivian Andes, the lines run dead straight for miles upon miles over mountains, through tunnels, across plains and salt flats where they crisscross each other, like seams in a huge patchwork quilt. Over two miles above sea level, they remain a mute witness to their forgotten purpose. To the west, down by the sea, near Nazca, Peru, the lines join gigantic geometric patterns and designs of familiar and strange creatures on Earth's surface. Unable to be seen from the ground, the patterns and designs were only discovered by people flying overhead in the last century. In Australia, Aborigines call their paths song lines and believe that in the distant past song lines covered the world.

Science knows energy waves are capricious as to where, when, and how they flow and are indifferent to straight lines. The Sun and Moon, sunspots and eclipses, the planets above and the topography below can influence them. Yet as British researcher John Mitchell observed, "the magnetic centers lie in straight rows across the country with a precision that characterizes human construction rather than the work of nature." In other words, man constructed the global web of dragon paths, however they are known and wherever they are found.

Yet no myths or traditions claim to have built the dragon paths and ley lines. When they mention them at all, they say someone else made them long, long ago. History books are silent about any potential candidates. Orthodox science ignores or dismisses the available evidence on who may have built the paths and lines and for what purpose, because it does not fit into the accepted worldview. Yet someone had a purpose and the skill to build them over the entire world. That effort created a legacy that has come down to us remembered only by straight lines in the earth, stone circles, and whispers of mystery and magic.

ARCHITECTURE

Man's hammer and chisel is evident everywhere in the ancient world. Creations of the past amaze us with their precise and aesthetic beauty, their massive stones, and their mystery.

East of Beirut in Lebanon wait the ruins at Baalbek. The uppermost portion of the temple is a sanctuary to the Roman gods Jupiter and Venus, but everything below that is an anomaly. At the base are rows of stone blocks 13 feet long weighing around 50 tons each. They are larger than the average stone blocks in Egypt's pyramids. On top of these rows, but still below the Roman stones, are three hewn limestone blocks called the Trilithon. Each measures 60 feet long, 13 feet high and 10 feet thick. Their individual weight can only be guessed: anywhere from 600 to 1,500 tons. They were raised 22 feet and placed perfectly on top of the base stones. No mortar was used. In fact a knife blade cannot be passed between them, so close is the fit. How these stones were transported from a nearby quarry is unknown.

Orthodox history files the ruin under Large Roman Temple and then ignores it. Yet someone was capable of designing the structure, cutting, dressing, and transporting the stones, then lifting them over 20 feet and placing them with perfect precision in their present positions. The Romans did not do it. Nor is there a construction company in the world today that can accomplish the same task. The age of these stones is unknown. Even the Romans thought them ancient. There is an Arab legend, though, that says the lower, original temple was built a short time after the Flood.

On the far side of the world, 11,000 feet high in the Peruvian Andes, lay the ruins of Sacsayhuaman, a fortress of massive stones in long parallel rows. The Spaniards believed it to be the Devil's work, so impressive was its strength. It served the Inca, but its builders remain a mystery. The gigantic stones used in the construction, some weighing up to 200 tons, are finely cut, some with up to 30 angles, and locked perfectly into place without the use of any mortar. Such precision is stunning, but more so is that these stone walls are earthquake proof. On a smaller scale, this style of architecture is evident throughout the Andes, and for uncounted years it has withstood earthquakes in a major quake zone. Whatever ancient engineers created this type of construction understood the region's geology, the dynamics of earthquakes and stone, and were masters at designing, cutting, transporting, and fitting immense stone slabs at high altitudes in landscapes filled with sky-touching mountains, plunging gorges, and roaring rivers.

Egypt's Giza pyramids have stunning precision in both their interior and exterior construction, so much so that if they were really tombs why was so much care taken when something less would have served a pharaoh just as well? Because they are something more. British engineer Christopher Dunn, author of *The Giza Power Plant,* writes, "Nobody does this kind of [precise] work unless there is a very high purpose for the artifact. . . . The only other reason that such precision would be created in an object would be that the tools that are used to create it are so precise that they are incapable of producing anything less than precision. With either scenario we are looking at a higher civilization in prehistory than what is currently accepted." There are those that argue that no tools capable of such precision have ever been found. This is true, but there is more evidence than just the missing tools. The pyramids, the finished products of those tools, and the intelligent minds behind them are the evidence.

A few hundred years ago, to enlarge the city of Cairo, the Great Pyramid at Giza was stripped of its casing stones, revealing the 2 million stone blocks underneath. Before that the Great Pyramid appeared as a wondrous, mathematically precise beacon of pure white, polished limestone. On nights of a full moon the pyramid illuminated the surrounding desert landscape in a soft glow and during the day blazed in solar glory as the sun passed overhead, reflecting light that, at times, could have been seen from the moon. This monument transcends our present worldview and our understanding because it is a synthesis of humanity and nature.

It achieves this because its proportions are harmonized. All living things from eggs to humans to trees embody harmonic proportions. This harmony emanates from nature and the universe. It takes an intuitive state of mind, with the appropriate knowledge and application skill, to re-create these expressions of nature in work done by human hands. Much ancient architecture, including many temples in Egypt and Greece and stone circles around the world, reflects in their design these proportions. Today, this ability, for the most part, has been lost. Proportion is like music. Music without harmony will sound harsh to the ear, and architecture without harmonic proportion will appear harsh to the eye. You can see and feel this in the barrenness of a strip mall or the hollow symbol of power in a skyscraper. Today, in our architecture and much of our science, we do not synthesize or harmonize, we analyze: We take apart everything with our nonemotional, analytic mind and thereby cannot see the forest for the trees.

MATHEMATICS

Another consistent intellectual achievement found worldwide in ancient architecture and writings, especially the most ancient, is a profound comprehension of numbers. Our ancestors understood that numbers reflected the structure of the universe they lived in. Numbers bring knowledge into the world. They are teachers, imbued with meaning. More than just abstractions, they can have universal, regional, and cultural significance. At a sporting event one team and its fans hold up their index fingers following a championship victory. A different meaning, more than just one finger, is understood. The number 13 is heavy with meaning beyond being the number that follows 12 and precedes 14.

For most of us, whole numbers, one, two, three and so on, serve us well enough throughout our lives. Each number, like everything, is made up of a multiplicity of smaller or larger portions: cells to molecules to atoms, or planets to solar systems to galaxies. These steps in either direction are infinite but meaningful because they are a part of the whole. Some of these proportions of numbers have been known from the earliest times. In the 5,000-year-old Hindu *Vedas* there are names for numbers as high as one trillion (1,000,000,000,000) and as small as three one hundred-millionths (0.00000003). The ancient Hindus, or Aryans, according to our history books, were cattle herders who raced around in horse-drawn chariots. What use would they have for such precise mathematics?

Two other interesting numbers are pi and phi. Pi is a specific proportion (3.1416 . . .) used to calculate the circumference, or external boundary, of any circle from its radius. Phi, also known as the Golden Section or Proportion, is a unique proportion (1.6180339 . . .) found by dividing a line in a way that the smaller portion is to the larger as the larger is to the whole.

For the majority of humanity today these numbers mean little or nothing. From a Beethoven symphony to the combustion engine to computer chips, the actual mechanics of these numbers are important to only a few. It is the end result that matters to the rest of us. Yet when we realize that these proportions are found continually throughout nature and the universe, these particular numbers become important. From a nautilus shell's beautiful, curving geometry to a spiral galaxy's long, graceful arms, and from the eight phases of a living cell's division to the fundamental octave of our musical scale, these "coincidental" proportions subtly underlie the world we live in. They are found in the images and writings of our myths and great religions. These proportions, when incorporated into architecture's length, width, height, and depth, create structures that reflect a delicate harmony, a sense of being a part of everything. It is a feeling that cannot be measured and counted, but one that humans respond to.

We are taught to believe that the Greeks discovered these mysterious numbers and proportions. But the Greeks admit that they went to school in Egypt, and we already know that the Egyptian civilization appeared "suddenly" in full bloom along the banks of the Nile. To add to the mystery, these numbers are also found within the ancient Mexican pyramids' dimensions in North America.

How were these numbers and proportions found in the first place? Would Stone Age wanderers who tended flocks of sheep or chased herds of mammoth or woolly rhinoceros be interested in 1.6180339 sheep or 3.1416 mammoths? How would they know what to look for?

Today we take these numbers for granted. Yet we did not discover them. These numbers appear only after a people reach a state of tranquility. Our ancient ancestors, 12,000 years ago, were people who saw similarities, not differences, between themselves and the natural world around them. This state can only come about in a society that has a harmonious relationship with its environment and has time to spend in inquisitive and creative ways. These numbers are like a lake with snowcapped mountains towering above. Its water, to refresh and illuminate humanity's innate curiosity, is deep and inexhaustible. From those depths these numbers were discovered, and then our ancient ancestors began to create and define an environment that reflected the harmony of the world.

In those numbers also lies buried an irony of our times. Mathematics, the science of numbers, is the only true universal language. Pi and phi, along with other mathematical constants such as prime numbers, are the same in every corner of the universe. The research program SETI, the Search for Extraterrestrial Intelligence, has great arrays of radio telescopes that send these universal numbers into space and listen for someone out there doing the same. Yet here on Earth, we have voices from our forgotten past communicating to us, in the only true universal language, and we are deaf and dumb to their message.

ASTRONOMY

Our planet's axis tilts 23.5°. This tilting is responsible for the seasons: spring, summer, fall, and winter. If Earth's axis were perfectly vertical our world would look very different. Her days and nights would be the same length, her climate would be much more temperate with the seasons being less dramatic than today, and her polar regions would be minimal. The abundant evidence we have found from 12,000 years ago describes such conditions.

Ancient Greek astronomers believed that during the Golden Age the stars rose from the equator in a horizontal position rather that the tilted one we have today. Japanese and Chinese traditions agree with this view. In England, Royal Astronomer Richard Proctor in 1873 commented that the aquatic constellations must have been named at a time when they lay "below the celestial equator," meaning in the sea or below the horizon, when the earth's axis was more upright and not at an angle. Studies by geophysicists, investigating the migration of the magnetic pole, suggest that 12,000 years ago, prior to what is called the Laschamp geomagnetic event, Earth's axis may have been orientated 30° different than today.

The Desana Indians of the Amazon remember such a time when the axis was straight up and down. Then something happened with suddenness and terrible violence, leaving uncounted dead and profound physical changes on the planet, including an altered axis. Ancient Egyptian and

Persian traditions refer to a time when the sea and the sky became "deranged." Eskimo myths in Greenland tell of a time when the Earth flipped over. Biological and geological evidence, as we have seen, also supports and reflects the cost of this derangement, and a cause that could only have come from an *external* source.

Because of the tilting and the eternal gravitational tug-of-war between the Sun, Moon, and Earth, our planet's axis wobbles, like a toy top. This changes the position of Earth's axis in relation to the stars above. For example, the northern end of the axis currently points at the star Polaris, also known as the pole star. In 12,000 years it will point at the star Vega, which will then assume the title of pole star. The northern axis will eventually return to its present location and point again at Polaris in 25,920 years, or the number of years it takes our wobbly axis to complete one revolution. This movement is known as precession. It is called precession because this rotation travels backward (clockwise) compared to the spinning of our planet (counterclockwise).

This precession is also evident at the spring equinox, March 21. At that time the Sun is seen to rise up against a background of a particular constellation of stars. For the past 2000 years it has been the constellation Pisces, and we have been living, during this time, in the Age of Pisces. In less than 200 years the Sun will begin rising in Aquarius. For some we are already entering what is called the Aquarian Age. For astronomers this celestial movement is called the precession of the equinox. This precession moves very slowly. It takes about 2,160 years to pass through one zodiacal constellation and, as previously mentioned, 25,920 years to complete one great cycle.

The Greek Hipparchus gets credit for first describing this precessional movement around 127 BC. Yet there appears to be evidence that it was known much earlier. Science historians Giorgio de Santillana and Herta von Dechend, authors of *Hamlet's Mill,* argue that our ancient ancestors had knowledge of this celestial cycle and incorporated it into many of humanity's earliest myths. They believe these myths to be of "awe-inspiring age," and that the precessional information's placement within them was deliberate.

Flowing against the current of mainstream thinking, their investigation uncovered a persistent "transmission" of recurring images, symbols, and numbers worldwide in mythic traditions. Beginning at a familiar point of departure, Shakespeare's Hamlet, Santillana and von Dechend follow the inspiration for that character back through time to Iceland and Finland, Rome and Iran, India and Polynesia. The now mythic individual is always associated with a mill of some sort that, for long ages, "ground out peace and plenty." At the heart of each mill is an axle, which Santillana and von Dechend believe represents the world's axis. Trees are also a persistent image representing the axle in a number of the myths, and sometimes both appear. The Maya and the Norse called it the World Tree. The Evenks of

Mongolia knew it as the Golden Post. What then occurs in all these myths is a terrible catastrophe. The axle breaks or the tree shatters, creating a potent image for a world once golden and now broken.

Precessional numbers in these myths and many others serve to crosscheck the validity of this theory. In Egyptian, Norse, Hebrew, Buddhist, Hindu, and Mayan myths and science, these numbers reappear, defying scholars to cry coincidence. These numbers all relate, in various permutations and regardless of the decimal's position or the number of zeros, to the sun's passage through the zodiacal constellations in the precession of the equinox. For example, 30 is the number of degrees allotted to each zodiacal constellation in the night sky. The number 72 refers to the years it takes the sun to pass through only 1 degree. Multiply 72 x 30 and you have 2,160, or the number of years it takes the sun to pass through only one of the 12 constellations of the zodiac. It takes 4,320 years to travel through any two of those constellations. We have yet to discover why the passage through two constellations is significant, but 4,320 consistently shows up across the world in myth.

At the end of the world in Norse mythology we find:

500 doors and 40 there are
I ween, in Valhalla's Walls
800 fighters through each door fare,
when to war with the Wolf they go.

British journalist Graham Hancock says, "With a lightness of touch that is almost subliminal, this verse encouraged us to count Valhalla's fighters, this momentarily obliging us to focus our attention on their total number." That number is 432,000.

According to early Jesuit scholars visiting China, the Imperial Library held among many things a massive work made up of 4,320 volumes. Within those volumes was a history of an ancient time when the planets changed their courses. The sun, moon, and stars altered their motions, Earth broke into pieces and the seas flowed over the planet. In Babylonian history the kings who ruled before the Flood reigned 432,000 years. The Kali Yuga, as the current Hindu epoch is known, contains 432,000 years.

Throughout these myths also appear members of the canine family: dogs, wolves, and jackals. Remember that those 432,000 fighters that issued from Valhalla went to fight a wolf. Santillana and von Dechend believe the canines are among a number of "markers" that help to alert those who listen to the myths that important information is found here.

Between the mills and trees and their apocalyptic destruction and axial derangements, numbers and members of the canine family, there are too many coincidences. Names and languages change, but the stories remain essentially the same. Santillana and von Dechend believe these stories reveal " a complex of uncommon images which nobody could claim had risen independently by spontaneous generation." They make it clear that this transmission of knowledge through myth is alerting us to something important. They acknowledge that much has been lost over the ages, and while we will never know the complete story, we can perceive fragments of this message. One thing we can tell from these fragments is whoever inserted these markers and precessional numbers into the myths were not primitive representatives of humanity, but members in an intelligent, advanced society.

Around the world ancient architectural sites stand as silent observatories revealing sunrises on the solstices and equinoxes, moonrise and moon settings, along with star, constellation, or planet alignments. For example, Avebury in England is a complex of sites that is manmade, blends seamlessly into the environment, and gives astronomical notice to a number of important days in the year, including a remarkable double sunrise from atop Silbury Hill. Machu Picchu, the mysterious sky city in the Peruvian Andes has the Intihuana or "hitching post of the sun." It is a single piece of curiously carved granite capable of measuring not only solstices and equinoxes but also lunar cycles. In Mexico, Teotihuacán may have had a population of 200,000 people long before the 8th century, making it larger than any city in Europe. The entire city was laid out in precise orientations not only for north-south alignments, but also as a mirror of the night sky above.

These astronomical functions, long suspected by some and long ignored by mainstream science, gave birth, in the last decades of the 20th century, to archeoastronomy. Lighting a fire under our arrogance in the 1960s, Scottish engineer Alexander Thom and British astronomer Gerald Hawkins proved that people with an intimate and complex knowledge of the heavens created Stonehenge, the great circle of stones in England, and many other megalithic structures from prehistoric Europe. Three decades later in *Uriel's Machine,* independent researchers Christopher Knight and Robert Lomas argue that these sites could also measure "the position of any bright object in the sky," changes in Earth's crust, and calculate the courses of comets.

Orthodox scientists point to megalomaniac pharaohs and superstitious blue-faced barbarians with bloodstained hands as the creators of these sites. And they give those colorful creators many reasons for the construction: fertility or death rituals, celebrations or sacrifices, or acknowledgment of Earth's natural cycles. What can be recognized are the obsessed curiosity they had with the sky and that many sites were built in aesthetic harmony with the surrounding landscape. But why would they want to track comets?

THE FLOOD

One of the most persistent and remarkable memories humanity shares is the Flood. So familiar to us from the book of Genesis, we do not realize it is just one of over 500 flood myths worldwide that describes a disaster or disasters so terrifying and destructive that the human race almost did not survive. These myths are uniquely similar, though differing in details.

Two basic flood myths exist. The first and most destructive is the mountain-topping flood. Many of these begin with a war in heaven, followed by a worldwide conflagration, fire and ice raining down from the sky, appalling storms that make hurricanes look like gentle spring showers, cataclysmic upheavals of land and sea, and a flood, in many instances boiling hot, roaring over Earth. They conclude with a time of such severe cold that rivers and lakes freeze and do not thaw in summer.

The second type of flood myth is one that swells up from the sea. A number of these floods follow a period of intense cold, and all describe how portions of the world sink below the rising waters. Humans and animals are sent fleeing in all directions. These floods are generally found alone in myths, unaccompanied by other disasters. As we will see, both these types of flood myths are correct.

Two overlooked features bolster this traditional testimony and help to validate the myths as enduring memories and not fictional stories. First, our ancient ancestors saw it coming, and they had time to broadcast a warning. This implies widespread astronomic observation skills, scientific knowledge, and a worldwide communication network 12,000 years ago.

We are familiar with Noah, who was warned by God,

but other gods advised their people as well. Prometheus warned Deucalion in Greece, the rain god Tlaloc warned the Aztecs, the gods of ancient Mesopotamia warned Utnapishtim, and the American Apache and Mojave Indians received similar warnings. Many were told to flee by animals: a fish in India, a llama warned the Inca, dogs (no doubt between fits of barking) told the Cherokee and Creek-Natchez Indians. After 12,000 years it does not matter if they were gods or talking animals that warned humanity. Our ancestors knew a disaster was about to befall Earth.

This brings us to the second validating feature. Our ancient ancestors understood the concept of gravity. Not just an apple-falls-from-the-tree understanding, but also a comprehension of gravity's power on a cosmic scale. Watching this object approach, our ancestors were aware of its course, size, and mass and believed they understood what the ramifications would be. They realized early on that boats would be their best chance, boats and arks that would hold not only themselves but also their animals and seeds for future crops. Why boats? If an object of sufficient mass was to pass close to Earth (it did not have to impact) and it possessed a powerful gravitational field (which humanity saw as Phaeton fought its way through the solar system), it would, with this field, create tides of tremendous size and force that could sweep over the entire planet. Boats and arks would be our best defense and protection.

Noah was not the only one who built an ark. Almost all

of our ancient ancestors who were warned built arks or found something that floated. H. Bellamy in *Moons, Myths and Man* reports "the Peruvians tell of whole fleets of arks." Implying shipbuilding and navigation skills 12,000 years ago, many of these arks, according to myths, were built far from the sea and high in the mountains. This is curious until we remember the builders' knowledge of gravity and tides and how the coming tides would be like no others ever seen before. Perhaps the mountains would give these last retreats of humanity precious extra minutes to complete their work. And as the waters rose to these heights and lifted the arks from their moorings, there would be less danger from thrashing the sides of mountains as there would if they were built in the valleys below. These were thoughtful considerations in a desperate time, but could not totally prepare the builders for the awesome event they would participate in.

THIRD MEDITATION

It is evening and we have rejoined the assembly. The public benches are filled and there is a sense of anticipation, electricity in the air. A bonfire is ablaze in the center to warm and illuminate. Flames stretch upward as smoke and sparks rise in spirals to the stars shining above. Dogs bark in the distance. The cranes, comfortable with our presence, chatter just beyond the fire's light.

We have been remembering our history, the present recalling the past, creating a connection, an understanding of where we come from and who we are. Our ancient ancestors, more than 12,000 years ago, lived in a world where night and day were of equal length, the climate was uniformly temperate, and plant and animal life was abundant. According to the myths, they lived in a world society founded on reverence for all life, a Golden Age. These myths are consistent in the events and many of the details they describe. They transcend time, distance, and cultural boundaries and are too numerous to have happened by chance. They are supported by evidence that formed the seeds from which our sciences, arts, medicines, and religions grew.

Then their world ended. Suddenly, brutally, and completely. The evidence of this worldwide disaster, as we have seen, is all around us if we know where to look. Plato even provides us with a written date for this terrible event: 9500 BC. Plato received his information by way of the Egyptians. Why would the Egyptians pass along that particular date? Did they make it up? If they did, it is one of the most amazing coincidences in history. Or was the date handed down year after year, century after century, millennium after millennium? It must have been a very important moment in time if so much effort was made to remember it.

When you take Plato's date and add it to the testimony gathered from our myths, evidence for an advanced society deep in antiquity, the enormous geological and biological evidence that we have seen is available, and put it all in the context of a planetary disaster 12,000 years ago, we find an internal consistency that science and its current worldview lack.

Science, for its part, has been misrepresenting or ignoring the evidence it has collected over the last 200 years. It has gathered an amazing repository of knowledge, but at times understands very little of it. Science is frozen in a sea of ice, a mile thick and millions of years old, unable or unwilling to move past its dogma and prejudices. It is one of the blindfolded men who are asked to describe an elephant. Grasping the trunk Science is sure it has the answer, but in reality it holds only a piece of the puzzle.

What has been recalled is a history unlike any we have been taught. Biology, geology, and our ignored human memories are a trinity, united in their testimony, which can bring about a new understanding between history and humanity. The scientific evidence written in stones and bones, and the indelible memories, cloaked in myth, put a human face on this catastrophe that allows us to remember our past with clarity. And memories, once awakened, have a way of surprising us all their own.

10

PHAETON AND EARTH

Nothing now remained between Phaeton and Earth. No doubt some people realized the crossing would take two to three days, but to most it did not matter. It took an eternity. Tremors of fear became deeper and more insistent. Social traditions became strained in the rising helplessness and foreboding, which now darkened everyone's thoughts.

The atmosphere crackled with burnt electricity. Fires grew in size. Plants and trees, shriveled and brittle in the blistering heat, offered no food or shade. Animals mad with thirst and hunger moved in frenzied masses. The Moon and Phaeton exchanged great arcing electromagnetic lightning bolts.

On Earth these arrows of light, followed moments later by startling booms, frayed people's nerves and magnified a sense of human powerlessness. Yet some maintained their focus and purpose, firm in their resolve to survive. Final supplies were loaded on many of the arks: seeds, animals, and tools. Other arks were already closed tight. Waiting.

When clouds sporadically parted, the night sky revealed a cosmic armada, radiant and grim, approaching Earth. The heavens were scarred with comets. Shooting stars appeared, burning up in the atmosphere. Phaeton, larger than the moon, flames spitting in all directions, plowed through waves of lunar and terrestrial magnetism, an almighty sun god dominating the heavens. His attendants, including Kingu, orbited around him like angels, seraphim bright and terrible. A trail of debris, resembling war banners glittering and dark, stretched back across space toward Mars. At times the agitated state of Earth's atmosphere caused Phaeton's appearance to change. The ancient Persians remembered how Phaeton transformed itself from "the shape of a man" to a "golden bull" to a "white horse with golden ears." Traditions around the world recall a dragon, thousand-headed, thousand-eyed, or horned fish, leviathans, feathered serpents, and monsters from heaven approaching Earth.

As when a dog prepares to lunge, thunder growled over the world. Then Earth shook as the first electromagnetic bolts shot toward Phaeton. The planet killer fired back, and Earth rocked from multiple impacts across her Northern and Southern Hemispheres. New fires sprang up and spread quickly, fed by the tinder-dry landscape and growing winds.

The ground quaked. As the gravitational fields of the combatants locked, Earth's spin began to slow, which sent winds howling. Earth's waters responded to Phaeton's call and started to flow north to the strongest point of gravitational attraction. Below the crust the viscous seas of magma moved north too. Earth started to wobble on its axis. Earthquakes multiplied in number, each more powerful than the preceding one.

Humanity's buildings and structures snapped like matchsticks, and crumbled like toy blocks. People and animals sought shelter where they could, sometimes huddled together in terrified groups. For some an inner realization opened the door to shock. They sat or stood with horrified gazes, each robbed of clarity of vision and presence of mind. Fear vibrated in their veins and then locked tightly into paralysis. They no longer reacted to the frantic pleas of family and friends.

Internal heat and pressure were building up under the Earth's crust. Magma in surges broke through and flowed in waves over the surface, incinerating everything it encountered. Traditions from the Americas, Europe, the Pacific Islands, and Asia describe humans and animals seeking refuge in water from the heat only to perish as the water itself heated to a boil. In Polynesian myth, the god Maui "jumped into the sea for protection, but the water was boiling with heat." The Voguls from the Ural Mountains describe in their myths that many were lost in the boiling waters. The Ami of Taiwan tell how the earth cracked open, and from fissures hot water flooded over the face of the world.

As on Earth, the Moon's crust was cracking from the power of Phaeton, magma exploding through and flowing over the lunar surface. Waves of debris began to bombard the Moon.

On Earth, new volcanoes rose quickly, spilling their steaming entrails. The planet rumbled with deep, resonant notes fading in and out of human hearing, then convulsed, its whole body shuddering and ringing like a great iron bell struck by an unrelenting hammer. Mountains crumbled in avalanches of earth and stone, obliterating all that stood in their way.

The death toll of plants and animals increased as cathedral-like forests burned and twisted in the winds, and skeletons of ruined cities disappeared under rock, lava, and flame. Once great and graceful herds of mammoth, impala, horse, and llama, broken by exhaustion, stopped, stood together, and perished. Seeking refuge in caves, people and animals were buried alive under collapsing mountains. As waters rose in the north, arks, on the lower hills, lifted from their moorings.

Phaeton and Earth continued to trade crippling bolts of lightning across the narrowing space between them. The atmosphere, choking with pollution and thick with cloud, sank to the ground as a dark curtain. Africa, Asia, and the Americas share the same memory: The sky fell and "almost all the people were killed." Amerindian tribes from the Pacific Northwest tell of a black veil so intense that "birds flew into one another and men and animals stumbled into one another." An age of darkness began.

Firestorms ignited into a worldwide conflagration. Flames, glowing in the darkness, swept over mountains and plains. Forests standing for millennia became cauldrons of shooting cinders and sparks, sending torrential funnels of superheated wind into the atmosphere. In their wakes air-breathing creatures suffocated on carbon monoxide and grit.

As Phaeton drew closer, Earth's rotation slowed almost to a halt. The winds, now a continuous roar, went berserk. Ancient Mesopotamians describe a "wind that had no equal." The North American Maya named this mighty wind *Hurakan*, our hurricane. But our hurricanes would shrink in comparison to these cyclonic monsters. Filled with dust, sand, and gravel from millions of tons of displaced topsoil, these punishing, heated winds raced over the horizons and around the world, wearing exposed stone smooth, carving others with gashes and channels, and sweeping away forests and hills. Huge boulders were picked up, as a child would lift a small toy, and heaved hundreds of miles.

Mastodons, sloths, great bison, birds of all species, humans, and vegetation, ripped from where they fell or hid, were sucked into these winds like pieces of straw. Brutal currents jerked victims back and forth across the sky, helpless puppets on invisible strings. Within the crumbled bodies of those discarded, external flesh held the smashed bones and organs together, to be buried with other debris and create the drift deposits around the world.

Earth's crust split and shifted under Phaeton's unrelenting pressure. Like giant furnace doors thrown open, the great fracture zones buried today at the bottom of the oceans split apart. Abysses opened around the globe, swallowing land and sea. Some closed in a gasp, others filled with migrating waters in vast cascading falls or with millions of tons of falling rock and soil. Fists of stone punched through the crust with horrifying sound and heat; their razor peaks stretching toward a sky streaked by lightning. From the bottoms of the seas, scalding heat and gases bubbled upward, lowering the water's density. Anything floating above lost buoyancy: plants, animals, people, and arks sank like stones. Across the landscape clouds of gases from deep within the planet burst to the surface. Some ignited into fireballs. The North American Zunis remember these heated blasts that brought the blood flowing in the veins of humans and animals to a boil. The blood then exploded from their bodies like magma from the earth. A primeval smell of destruction and creation shrouded the planet.

Life was overwhelmed, everything dying in great numbers. Yet others, still clinging to life, were hiding in arks on the heaving waters, in caves or hidden canyons overlooked by the chaos sweeping over them.

The Southern Hemisphere drained of water as the oceans and massive rain clouds followed the pull of Phaeton's and Kingu's gravity fields north. The northern skies darkened above, and the waters climbed upon one another, rising like a mountain into the sky. This water mountain was held fast by Phaeton's cosmic grip. From Lapland, Sami tradition recalls the seas gathered "up into a huge towering wall." Land in the Northern Hemisphere disappeared.

Arks built on the higher slopes now floated free upon the waters. Lions, wooly rhinos, goats, snakes, and people fled up hills and mountains. Others of countless species, drenched by the water's blinding spray, dodged the living and the dead, great and small, carried by the raging wind. They heard the cries of their kin and climbed higher to the last pieces of solid land.

The combination of Phaeton's and Kingu's gravitational fields and the immense weight of the water in the north pulled the planet over. Eskimos in Greenland speak of a distant time when the Earth "turned over." The ancestors of the Pawnee of North America remember the North Pole

Star changing places with the South Pole Star. Earth righted itself, and then the axis tilted, more than 20 degrees, under the staggering violence of this cosmic war. Thunder roared. Earthquakes continued to roll over the world. As Phaeton and Kingu drew closer, Earth and the Moon fought back, pulling Kingu away from its dreadful master. Entering into Earth's Roche field, Kingu began to break apart.

Darkness and storms reigned in the north, broken by blinding bolts of electromagnetic discharges and the glow and vomit of volcanoes. Lava in rivers and streams of gold, silver, and copper surged across the landscape, sometimes one atop another in ribbonlike seams. With the invader pulling many of the rain-filled storm clouds north, the sky and the heavens in the Southern Hemisphere revealed themselves through the tormented atmosphere. And to the terrified and exhausted survivors a new tribulation appeared: Phaeton's host had arrived.

Like some terrible nightmare this awful spectacle closed in on Earth: flaming star matter, comets, asteroids, meteors, a disintegrating Kingu, and cosmic dust swirling in vast eddies and sparkling whirlpools. Swarming like a plague of insects over the Moon, the stone, metal, ice, and long ribbons of gas that had followed Phaeton across space now began their descent and bombardment of Earth.

As they plunged into the terrestrial atmosphere, Kingu, behind them, was torn apart by Earth's Roche field. Amid a crashing din, streams of liquid spurted across the electrically charged space, ignited and fell toward Earth in rivers of fire. Ice from Tiamat's and Kingu's atmospheres and Mars's stolen oceans, sparkling in Phaeton's reflected light, whistled and howled down into the darkness, forming into armies of giant hailstones. Fiery meteors flanked their descent. Some disintegrated in blinding flashes, while others smashed into the surface below.

Kingu's disintegrating body now exploded into millions of pieces. Fragments ricocheted off Earth's atmosphere while others rocketed into it as an advancing wall of stone and fire. Fireballs rose up from the dark horizon and then roared overhead. They pounded the Earth, creating tens of thousands of parallel and elliptical craters. In biblical tradition, a great star called Wormwood was seen falling in flames to Earth. The Cashinaua Indians of Brazil remember how "heaven burst and fragments fell down and killed everything and everybody." No army ever faced such an invader. In all our mythologies very few of our ancient gods stood up to this assault. Most cowered in deep places or had fled long before. Zeus was one who stood and hurled thunderbolts at Phaeton, but it was in the Norse myths where the gods made the most heroic stand. At Ragnarok, the twilight of the gods, the 432,000 warriors sprang from Valhalla's doors to defend Earth. One-eyed Odin faced Phaeton's attendants, the Sons of Muspelheim, and died fighting a great wolf. And mighty Thor, with shield held to high port and hammer swinging, fell.

In the north under the heaped-up waters, whales, dolphins, and other great schools of fish stayed deep as their world agitated and in places boiled. Events on land were multiplying madly, incompatible with hope. Entire herds, whole families, grandparents, parents, and children, of humans, mammoths, and monkeys, gazed for a final time into one another's eyes and died.

When this first meteor wave crashed through Earth's heavy clouds, the moisture building from weeks of evaporation burst. Heaven's window shattered, and waves of rain fell in a deluge.

With Earth's rain fell alien rain: The Iranian scripture, Zend-Avesta, says drops were the size of a man's head and sometimes boiling hot. The North American Sac and Fox peoples tell of raindrops the size of a bull's head or the size of a wigwam. Many of the planetwide firestorms were extinguished. In the north, the water mountain still held fast in Phaeton's grip swallowed millions of tons of the cosmic rain.

There was no respite. Everything seemed to exist in a universe of painful sound. Earth, palpitating with earthquakes, continued to exchange energy blasts with Phaeton. Hail and stone storms lashed the land and sea. Wave after wave of iron meteors leveled forests and embedded themselves into hilltops and mountains. The ancient Hittites called iron "the fire from heaven." The Hebrews name for iron was *nechoshet,* droppings of the serpent, Satan. Kingu's burning gases and liquids fire-danced across Earth's skies, the gases evaporating, the liquids raining to the ground in sticky globs burying plants and animals. The Klamath Indians described it as a "rain of burning pitch."

Red and brown oxidized metals from Tiamat, vaporized into dust at her death, now mixed with Kingu's fluids and fell in great clouds and streams over the world. In Finland it was called "red milk." The Tartars remember a "rain of blood." The Bible recalls "hail and fire mingled with blood" falling from the heavens. From Tiamat's broken core, nickel-rich magnetic spherules and radioactive nodules, the size of peas, the size of bricks, fell in monsoon-like waves over the planet.

In the north, upon a towering, drunken sea, the arks floated. How many were lost to wind, waves, and meteors, disappearing forever under the waves? Within those fragile walls were our past and hope of a future, helpless. Waiting. In an alliance of soul-shattering trauma, humans and animals around the world, hidden in caves and wherever else they found shelter, waited.

At last Phaeton moved away from Earth, and his once unyielding grip lessened on the water mountain in the north.

First the waters moved slowly. A Navaho myth says that when it appeared the waters filled the horizon except in the west, advancing like a chain of mountains. The Pima Indians remember how "a green-water mountain rose over the plain. For a short time it seemed to stand upright like a wall. Then it was split by a vivid flash of lightning and plunged forward like a ravenous beast."

Amid unrelenting rain, darkness that touched the ground, and flashes of lightning and streaking meteorites, the Flood, in titanic chasms of roaring water, poured over the world.

Beneath Earth's crust the seas of magma were freed as well and raced south, toward the equator, in huge undulating waves. The planet's crust rose and fell like a rolling ocean. The Himalayas, Rockies, and Alps rose to new mountainous heights. In the far north, with the magma's sudden departure, the surface collapsed into an immense basin, disintegrating islands and continents while capturing an ocean that filled it from towering cataracts.

The Flood ripped open Earth's surface, gouging out rock and earth, creating canyons and valleys by the thousands, while filling others with mountains of debris. It rammed boulders into stone mountains and scoured other mountains clean. Waves roared up heights inscribing watermarks thousands of feet high, depositing tons of debris and stones the size of ships along mountain ridge lines.

Carried along helplessly, through driving rain and atop this plunging sea without a shore, were humanity's arks and gigantic rafts of vegetation and smashed forests with birds, animals, and humans clinging for life within them.

The wooden arks proved vulnerable: Their hollow compartments, once breached, filled quickly and sank, dooming their inhabitants. Others capsized, unable to withstand such leviathan roller-coaster waves. But arks made of densely packed reeds bobbed like corks, even when waves washed over them.

Like a gigantic aqueous creature without hands or feet, bones or blood, the Flood rushed on, bellowing with a roar so deep and bottomless it bludgeoned the senses.

On land, families and strangers, humans and animals who sought shelter from fire and bombardment in underground havens were killed instantly when iron fists of water penetrated their refuges, crushing their remains into the smallest and deepest crevices. At times it seemed the mountains themselves were sailing like out-of-control ships on an insane sea. Survivors, high on the slopes, clung desperately to one another and anything else they could grip. Swells rose and fell with the velocity of a bullet. One moment groups of humans and animals huddled together, and the next moment they were swept away, gone forever.

The duration of Phaeton's onslaught can only be guessed: perhaps 48 hours, perhaps a few days. In time the bombardment lessened, and in its own time the Flood subsided, finding haven in the vast depressions of a newly sculpted Earth. Arks, rafts, and islands of flotsam now floated on new seas or rested upon fresh shores. Left behind were saturated landmasses, muck beds, and debris fields filled with layers of volcanic ash, gravels and clays, vegetation and dismembered remains of countless animals, some to be found 12,000 years later in caves 1,000 feet deep and at elevations more than 2,000 feet above sea level. Despite the calming of the waters, intense rains still fell, lightning flashed, meteorites lit up across the blackened skies, and the ground trembled.

Phaeton in his arrogance, in the guise of ancient Mesopotamia's Marduk, boasted of this time:

> *When I stood up from my seat and let the flood break in,*
> *Then the judgment of Earth and Heaven went out of joint . . .*
> *The gods, which trembled, the stars of heaven—*
> *Their positions changed, and I did not bring them back.*

The planet killer moved deeper into the solar system. Yet with Phaeton's passing, on a cloud-covered world, punished by drought, fire, meteors, rain, and wave, the final hammer blow was about to fall.

Brought on by a sunless, polluted atmosphere, massive precipitation and plunging temperatures, winter arrived. Frost and ice, twin arrows impaling an already wounded body, raced outward from newborn polar regions. Trapped water in canyons, craters, and depressions at high elevations or latitudes froze over. The muck beds in Alaska, with their uncountable shattered bodies, and the mammoths and other animal carcasses strung out along northern Siberia froze. Eight thousand years later, the Old Testament would ask, "Out of whose womb came forth the ice? And the hoary frost of heaven, who hath gendered it? The waters are hid as with a stone, and the face of the deep is frozen." The Norse would call it the Fimbul Winter. The Iron Winter. Twelve thousand years ago, Earth's Ice Age, a child of both appalling heat and cold, was born.

As this birth occurred, Phaeton sped toward the Sun, whose grip was now inescapable. A dwindling but still formidable host of comets, meteors, and debris continued to follow their master. They approached Venus and fought one last battle, obscured by clouds and, at the moment, of little concern to the survivors on Earth.

Today Venus's scars reveal a terrible struggle on the doorstep of the Sun. The second planet's surface is geologically very young, a result of catastrophic changes. Lava, like a great coat, blankets much of the planet. Active volcanoes still rise from a land disfigured by impact craters and crater clusters, where large meteorites broke apart seconds before smashing into the planet. Unlike Earth, whose atmospheric conditions were bringing on an ice age, Venus's dense, cloud-filled atmosphere contributed intense heat and today reflects back more light into space than any planet in the solar system. Phaeton's inconceivable power is evident in the inclination of Venus's axis. It is wrenched over a staggering 177 degrees. She is almost upside down resulting in a rotation opposite to that of the other planets. On Venus the Sun rises in the west and sets in the east.

Everything, whether living or not, has a beginning and an end. Phaeton, born in a supernova, left a convulsing Venus behind and moved forward to meet its destiny. There the Sun, patiently and irresistibly, drew Phaeton into its embrace, and there it swallowed the little sun whole, like a bullfrog gulping down an insignificant insect on a sunny day.

11

AFTER THE FLOOD

THE HEROES

After the Flood subsided and the waters rested in their new basins, Utnapishtim, the Mesopotamian inspiration for Noah, opened a window on his ark and looked out upon the world. Everything was the color of mud. Corpses, human and animal, thick as seaweed floated on the water. Utnapishtim sat down and wept.

A new world now existed, born at an incomprehensible cost. Massive new continents, *our* continents, stretched mostly north and south, instead of east and west as they had a few short days ago. New oceans and seas, *our* oceans and seas, broke their waves on virgin shores. Fresh-faced mountains reached toward the skies, some continuing to rise while others still glowed red from their ardent and violent birth. In North America Zuni myths describe that, after the disaster, earthquakes still shook a world covered by "vast plains of dust, ashes and cinders, reddened as in the mud of a hearth-place."

Some arks found land quickly. Utnapishtim made landfall in seven days. South American myths tell of survivors, high in the Andes, seeing five arks resting on the slope of a mountain after the waters had retreated. Out of one emerged Paricaca, a cultural hero who taught irrigation to the survivors. Native North American mythology records how western mountains became resting places for arks following the disaster. The Namaqua people together with their cattle arrived in an ark on the southern African coast and settled there.

Other arks were not as fortunate. Noah floundered on the waters for five months before coming to rest on Mount Ararat. Traditions tell of drifting on unknown seas, an ordeal racked by anxiety and fear of ever finding land again, if there was any land left. No one aboard the arks knew where they were. Navigation by the sun and the stars was impossible. Much of the atmosphere was heavily polluted, and even if the stars were occasionally glimpsed, their positions—or rather Earth's position—had changed with the tilting of her axis. These survivors of Phaeton's visit were doomed to float on newly forming currents or to be lost in the doldrums, listless and despondent on calm seas one moment, then racked by squalls the next. A few never touched land again. On *our* Pacific and Atlantic oceans floated ghost ships, their human, animal, and plant cargo often perishing before land was reached.

Many arks did make landfall, beached alongside monstrous rafts. Wading ashore, the survivors, whipped by the stench of decomposing flesh and bones, saw fish, whales, birds, humans and, uncounted animals great and small, all mangled and broken within a wooden maze of shattered trees, rotting vegetation, and seaweed.

Earth's higher latitudes, particularly in the northlands, felt the lash of freezing rain. With the dark skies and the planet's altered axis, temperatures grew colder and colder, and exhausted survivors found themselves in nearly hopeless situations.

Volcanoes continued to erupt across the world, producing heat that evaporated the waters that condensed into vapor, which fell as rain and snow. Fog, cooled by winds and abetted by freezing drizzle and rain, formed clusters of ice and rime, layer upon layer upon layer. The Yamaha Indians of the Andes recall that after the Flood "a coat of snow and ice covered the whole country." Stories of the Tongans in the South Pacific tell of *Taifatu,* the ice-covered ocean.

Survivors on mountaintops stood and gazed out over a profound desolation and wondered if they were the last of their race. Hunger and thirst drove some, in spite of their fear, down from these heights of salvation. Others stood next to their arks and did not leave the mountaintop. Those arks became the first hearths on a new Earth.

Across much of the planet layers of silt, loam, and volcanic and cosmic ash coated valleys and hills and served as unmarked tombs for the dead and as future nurseries for seeds of plants now gone. Cooling seams of gold, silver, and copper lacerated a tortured surface, offering the only color, like bright ribbons, when a brief flash of sunlight here or a moonbeam there reflected off of them. Iron mountains sat motionless after their interplanetary journey or their scalding uplift from deep within the planet.

In the mythology of Africa and the Americas, caves are recurring locations for the birth of humanity. The Navajos of North America remember how they, along with other tribes, including white people and animals, were sealed up together in caves. For food they ate some of the animals trapped with them. It took days to dig their way out, and when they broke through to the surface they found themselves on a mountain surrounded by water. In time the water flowed away, leaving a world of mud. There was minimal light even during the day "for there was yet no heaven, no sun, no moon, nor stars."

A cultural hero of Brazil's Karaya Indians, Kaboi, helped dig out his people after the Flood. To the west, across the new divide of the Andes, the Peruvians had a name for caves: *paccarisscas* that referred to "places where their ancestors originally emerged" into the world. In Asia, hill tribes of Bangladesh recall a great chief, Tlandrokpah, who lead them out from caverns. Heroes were not limited to humans. To a number of North American tribes, the rabbit burrowed its way first, followed by humanity, into the new world.

Out of sight, but never out of mind, these images of ark, mountain, and cave strike deep chords within us. Throughout history they have been powerful symbols with hidden and lurking presences: life preserving, terrifying, holy, mysterious, towers of strength and sanctuaries for humanity in our darkest hours. These imprints on our memories first took shape in the days immediately following Phaeton's departure.

The Zuni remember other places of safety: "dark canyons, deep valleys, sunken plains," that somehow were overlooked by the fires, halted perhaps in their tracks by waves of howling winds blowing against them, or the Flood swerving at the last moment in another direction. Trees still grew there, "and grasses, and even flowers continued to bloom." But these lucky canyons and last valleys were exceptions rather than the rule.

For the vast majority of living organisms survival was at once simple and complex. Severe trauma, worn like heavy shackles, hindered every attempt at recovery. Burdened with overwhelming fatigue, mental and physical exhaustion, our ancestors tried to rest. Some, with souls shattered and haunted, fell still and silent and did not survive. Others shook with fear, while anger, regret, and helplessness pulsed like a fever in their minds. Those who slept did not escape the pain, whether held in thrall by dreams of those now gone, their heart-wrenching partings relived over and over, or by aching fear exploding into nightmares. Though they might have recalled the mechanics of the disaster, the scale of the horror they experienced was unfathomable, and when they tried to look into the future with its sunless skies and its cold, incomprehension must have etched itself like dagger slashes across their faces.

Other people, however, came forward. They were the unknown heroes whose penetrating influence created hope in the broken survivors. These men and women held onto their gentleness, strength, and clarity. They knew what needed to be done, and ceaselessly, like wind dispersing clouds, reached deep into fear and darkness and brought back human beings.

Success, at this point in time, could only be measured in small increments. Basic needs had to be met first, and care given to those who hung on to life or sanity by the barest of threads. Food, water, and warmth were the most urgent requirements. The catastrophe contaminated or destroyed much of Earth's fresh water. Even rainwater falling through horribly polluted skies was, at first, only marginally safe. Many streams were rust colored, bitter, and even poisonous, from banks washed with carcasses, slime and dreck, and wastes of liquefied and oxidized remnants of Tiamat and Kingu. When the biblical star Wormwood (most likely pieces of Kingu, or perhaps one of Phaeton's satellite "whirlwinds") "fell from the sky, burning like a ball of fire," and crashed into Earth during the bombardment, its chemical composition made the water undrinkable. The Bible remembers, "many men died" drinking the bitter Wormwood water.

Some survivors had stored their own water in arks and caves, but those without it soon found themselves desperate. Any discovery of safe, fresh water in a stream, a bubbling spring or a well made a deep impression. Folklore from around the world is rich in accounts of life-giving wells and springs, and the symbolism that comes down to us of rebirth, healing, and baptism is a reflection of those initial life-saving discoveries.

Starvation constantly lurked and the search for food became a desperate act. The Aztecs remember how "with great toil and weariness" people traveled each day "going by mountain and wilderness seeking their food; so faint and enfeebled are they that their bowels cleave to their ribs . . . [their] face and body in likeness of death." The Chinese wrote that after the disaster their ancestors "dwelt in caves and desert places, eating raw flesh and drinking blood."

The Maya tell of "the great pain [the people] went through; there was nothing to eat, nothing to feed on." Babylonian myths relate that after the Flood, "the womb of the earth did not bear . . . vegetation did not sprout . . . the broad plain was choked with salt." The abundance of rotting vegetation and flesh provided food for a short time, but food poisoning took a terrible toll in lives. Survivors who were vegetarians before Phaeton found no recourse: Meat was the only nourishment. Animal and human meat.

Difficult choices were made for survival during those first days.

Throughout the first year following Phaeton, an enormous diaspora took place across the planet. Vulnerable to disease, malnutrition, and predation, decimated herds of animals and humanity became wanderers. Like fire on a mountain, most did not linger in any one place. They stripped land of anything edible and then continued on in a ceaseless search for food and water, warmth and shelter. The Mayan text, the *Popol Vuh,* remembers how "there came to be many people in the blackness; they began to abound even before the birth of the sun and the light . . . they did not know where they were going." Beside the Mayans themselves, there were "black people, white people, people of many faces," traveling in all directions.

Many people and animals abandoned the northlands and higher elevations because of the unrelenting winter. Science refers to this time, 12,000 years ago, as the Younger Dryas—a thousand-year period of severe cold that held the high latitudes in a frozen grip. Storms blew hard and cold, with heavy snowfalls and mighty frosts. The Norse "children hardly kept alive in that dread winter." The Maya, farther south, recall in their *Popol Vuh* "white hail and blackening storms. The cold was incalculable."

As they took those first steps over a new Earth, few could feel the vast wonder of creation, the new world of possibility before them. Most found it unbearable to look upon the stark loneliness, the darkness, and the cold.

Heaps of unburied dead lay along the beaches, resurrected and delivered from the Flood by new currents and tides. Bloated and rotting, the stench was horrific. The survivors found deep wounds inflicted on their Earth. They knew they would never again see forests at dawn flowing to the horizon like a wave on the water, rolling plains of grass and golden grain, blooming gardens wet with a gentle rain, the laughter and the smiles of family and friends, or a life without fear. Never again would they see immense herds of wooly rhino, mastodon, and camel, the sky filled with birds, or hear the joyful first trumpeting of young mammoths in the spring. It was all gone forever.

Remnants of Phaeton's host returned after a year had passed. Though much of the cosmic horde perished plunging into the Sun with their master, our system's planetary gravitational fields captured uncounted comets, asteroids, and meteors that took up eccentric orbits throughout the system. On Earth, when the clouds broke, comets could again be seen. Hearts chilled and people panicked. Meteors of stone, iron, and ice came out of nowhere. Hundreds of thousands per hour: streaks of light racing across the sky, fireballs exploding over their heads. People and animals fled again to caves or high ground.

Though random and regional, and not a global barrage, these showers of fire and stone succeeded in reinforcing the trauma of the catastrophe, burying insecurity and dread deeper into our ancestor's hearts and minds. And over the long, cold years that followed, these recurring bombardments contributed to the belief that humanity was no longer a part of nature. Nature had turned on them with viciousness, taunting them year after year until our ancestors began to ask why. And every time the showers of missiles returned, the unbearable memories would too. There had been no time for burials or grieving and no ceremonies to ease the pain. The only remembrance was through a kaleidoscope of nightmares that haunted each dark day.

Though the volcanism that caused the polluted skies was lessening, this dark age may have lasted up to 100 years. Surely nature had become our enemy.

The decades following the loss of the old world were extremely difficult for plant and animal life on Earth as well as humans. Many plant species disappeared forever, vanishing during Phaeton's assault or from the dynamic climate changes that followed. But the seeds of other plants lay buried in soil, surviving until conditions improved. The decomposing plants and animals and the mineral wealth hidden in the layers of volcanic and cosmic ash encouraged a rebirth of plant life. In the polar regions it failed as glaciers spread down from the mountains and the land hardened into permafrost, but in the more temperate and tropical lands life rebounded, as a result of the altered axis and perhaps more sunlight and heat. Grasses appeared first, and the Babylonians tell of how people had only grass to eat for a year. Shrubs followed, and then slowly over the years trees began to stretch to the skies.

For the animals the task was as difficult as the one that

faced humanity. Some species, particularly the larger mammals—mammoth, mastodon, and giant ground sloth—moved toward extinction. Most likely multiple factors, including the loss of temperate lands, the once abundant food supply, the shocking numbers who perished in the disaster, their weakened physical conditions, and now the relentless hunting by starving humans, doomed them forever.

THE TEACHERS

In the wake of Phaeton's passage through our solar system and the subsequent horrific environmental disaster that overwhelmed Earth and all its inhabitants, many of our ancient ancestors developed personality disorders. Today, 12,000 years later, we recognize these as post-traumatic stress disorder. How can we know this? We can only speculate, but it is speculation based on what we know of human responses to lesser traumas today. From the sudden unexpected death of a loved one to destruction and the deaths of thousands or millions at the hands of madmen or natural disasters, shock and trauma are symptoms born of these events.

For survivors of Phaeton's violation of our planet, severe emotional shock produced deep psychic numbing, depression and anxiety, and denial that in some cases disappeared into amnesia. Episodes of rage and abuse likely erupted, as well as flashback experiences, insomnia, suicide, and survivor's guilt. Children were particularly vulnerable, revealing signs of damage ranging from apathy to aggression. The newborn entering the scarred world after Phaeton received childrearing grounded in fear and uncertainty, rather than the reverence and security once found in the world before the catastrophe.

During the first 100 years following Phaeton, the spiritual tradition that bound humanity together with their world also disintegrated. As generations followed one another, the spirit of the Golden Age faded and, in places, disappeared. People's faith was lost. Once, in that wondrous age, our ancient ancestors felt a part of everything, but now people were distant, detached, and fearful. Their physical and psychological struggle for survival separated them from nature and from one another. Survival was paramount. There was little guidance. Over the decades some forgot the One Law of doing unto others as you would have them do unto you. Others acquired a taste for human flesh. The drastically changed climate, the devastating earthquakes and tidal waves that still unexpectedly passed over the world as she adjusted to her wounds, the unpredictable reappearances of Phaeton's host, the growing sense of utter insignificance and helplessness in the universe, and the gnawing fear that there would not be enough food to survive another day drove humanity to a dangerous crossroad.

Then one day, like a piper at dawn's gate ringing in a new age, the Sun rose up from the horizon. Few of the original survivors witnessed this dreamed-of moment. It was their children or their children's children, those who grew up in the post-Phaeton world of gloom and windy half-light, who saw that first clear sunrise. Many creation myths recount how, over the long years, the darkness slowly gave way to shades of muddy brown where no division between land, sea, and sky could be discerned. But as the atmospheric pollution lessened, the horizon became visible: The land separated from the sea, sea separated from the sky, and the sky rose to meet the heavens above. And at night the silver Moon and brilliant planets and millions of stars gazed quietly back.

While most of humanity now lived a stone-age existence, there were enclaves of our ancestors who came through the disaster with the seeds and animals, tools and knowledge they had stored away on arks, in caves, or on mountaintops. They maintained the strong holistic worldview from the Golden Age. They understood that their precious animals and seeds were not to be eaten: They were the future, the building blocks of tomorrow's herds and fields, the security from starvation.

With the return of the Sun, Moon and stars, a plan was implemented, created in the last desperate days before the disaster or put together over the years following it. Individuals, from at least one center of preserved knowledge, went out to explore and map the new world and to find and assist the lost people of Earth.

This remarkable theme runs through the world's myths.

The spirit of the Golden Age, revealed in these stories, emanated from these individuals, who reached out with all the practical knowledge and wisdom they had, to everyone they could find. Traveling by ship or foot across the width and breadth of Earth, these heroes and teachers arrived both alone and with companions. According to the myths, they tended to be tall, bearded, fair of skin, and wore long robes. They practiced what they preached; the One Law was a recurring theme. More than one myth refers to their gentle ways of persuasion. Replacing depression with courage, doubt with certainty, and despair with hope, they were the wood that, once kindled, allowed the lost spirit to return and burn as a fire in the hearts of humanity.

Some myths suggest these heroes and teachers appeared immediately after the Flood, and a number of them, like Kaboi in South America, assisted in rescuing survivors right after the disaster. However, for years after Phaeton, while everything was in chaos, immediate needs had to be addressed first, and the decades of darkness and unstable weather conditions made both navigation on the seas and travel overland difficult and dangerous. The people these later heroes and teachers found had, according to the Babylonian Berossus, a "wretched existence and lived without rule after the manner of beasts." Many were naked, living in caves, and had lost most if not all of the knowledge of the Golden Age, including how to make fire. To the children or grandchildren of the original survivors, the disaster was history, and tales of a Golden Age were fantasy. They had nothing in their experience to give meaning or reality to the events and wonders described to them. Now, finally, with the skies clearing, the time was right for those who had been luckier to try and restore something of what had been lost. Humanity's future was at stake.

Prometheus is perhaps the most well known of these heroes and teachers. In Greek mythology he is known as a protector and benefactor to humanity. He brought fire, agriculture, and architecture. In some accounts he even helped to create man. In ancient Mesopotamian myths he is called Oannes. Arriving by sea he gave humanity "the use of letters, sciences and arts . . . he taught how to construct cities . . . to compile laws and explained the principles of geometrical knowledge . . . he made them distinguish the seeds of the earth, and showed them how to collect the fruits." Oannes eventually departed, but from time to time others like him would appear.

Ancient Egypt had Osiris, who began his work by abolishing cannibalism and reintroducing use of domesticated grains. He and his companions also shared their engineering knowledge, particularly of canals and other systems of irrigation. The Greek historian Diodorus Siculus wrote how Osiris traveled south to Ethiopia, then across to Arabia, and on to India with his knowledge and wisdom, and that he urged mankind to "give up their savagery and adopt a gentle manner of life."

Teachers also appeared in the Americas. Quetzalcoatl and Kukulcan visited Central America. According to the myths both traveled with companions, stopped cannibalism, introduced corn and agriculture and the domestication of animals. They both shared medical knowledge, such as curing blindness and infertility, and taught mathematics, astronomy, architecture, music, and art. They introduced the calendar. They were said to have measured the planet during their journeys. Both brought peace and "taught that no living thing was to be harmed." Kukulcan is said to have stayed ten years in the Yucatán region before sailing off into the eastern sun. Quetzalcoatl, who also introduced fire to the people of central Mexico, is said to have departed after a time to the east on a boat made of serpents and birds.

Far to the south, amid the towering Andes Mountains of Peru and Bolivia, Con Ticci Viracocha is the hero's name found in myths. He too came with companions, shared medical knowledge, and revealed the secrets of domestication and agriculture, architecture and the engineering behind those ancient and amazingly durable earthquake-proof walls found in this region. He is also credited with the building of the megalithic city Tiahuanaco more than two miles above sea level. He taught the people writing, but that skill was lost again before the Incas (who acknowledge this fact) arrived thousands of years later. Unlike Quetzalcoatl and Kukulcan who both returned east, Con Ticci Viracocha departed across the Pacific Ocean. In the Marquesas Islands, 4,000 miles to the west, the Polynesians remember Tiki, a cultural hero who came from the rising sun, and was the first to populate the islands. Viracocha is also credited with introducing the style of agricultural terracing that covers the flanks of the Andes Mountains even to this day. This identical style of terracing is found across the Pacific from Japan and China to the Philippines and Indonesia as well.

As they taught and shared the remnants of Golden Age knowledge, the teachers introduced information they believed important to transmit to future generations. Many of our myths were created during this time. The local populations shared the story of the terrible disaster with their children and grandchildren, and the teachers shared something more. In the earliest worldwide myths, according to de Santillana and von Dechend in *Hamlet's Mill,* a common element was the precession of the equinoxes, represented by the great grinding-mill turning out peace and plentitude until it is shattered in a catastrophe. The mythic imagery was ambiguous enough: innocent and entertaining. Yet the scientific data and historical implications were accessible if the clues, the special numbers and barking dogs, for example, could be understood.

According to tradition Osiris, when he finished his teaching, returned to Egypt. Was this one of the enclaves of Golden Age knowledge that the heroes and teachers came from? Did Quetzalcoatl and Kukulcan return there, after they departed from the Americas? Following the disaster 12,000 years ago and the slow environmental recovery of the planet, Egypt was a land of temperate climate, blessed with adequate moisture, grasslands (which would not become a desert until thousands of years in the future), and a river, flowing from deep within a new continent's interior, that broke up into a many-fingered delta as it approached the sea. At the apex of this delta, the survivors settled. This piece of land is also at the center of Earth's landmass. From this point, if you drew a line on a world map east and west, and then north and south around the globe, these lines cross more land than any other.

If this site was found by accident it was a most extraordinary coincidence, but if it was found by design it could only have occurred *after* the cataclysm and the rearrangement of the continents. And it could *only* be found by survivors who had retained the sciences of geography and mathematics, and who had the appropriate navigation skills to explore and map the new world. Recall that many of the ancient Portolan maps that researcher Charles Hapgood studied had their original projections centered in Egypt.

Also at this apex sit the pyramids of Giza. Is their location on this site another extraordinary coincidence or is it by design?

When Osiris returned to Egypt he found his companions had been busy. They had begun building a new center for humanity. Another name that comes down to us from this time is Thoth. The Greeks call him Hermes, and traditionally he is credited with being the patron of the sciences: astronomy, botany, and mathematics. He is said to have measured the Earth and championed writing. He is also credited in many legends, particularly Arab and

Jewish, with the design and construction of the Great Pyramid.

A perpetual circus of research and speculation surrounds the pyramids at Giza. Alternative ideas abound teeming with clever, serious, and absurd notions about the when, why, and who behind the monuments. Egyptologists, depending on their personal chronologies, believe that the Great Pyramid was constructed between 2500 and 3000 BC. The most recent date requires all necessary technology be developed only a few hundred years after emerging from the Stone Age, which would be a remarkable feat. The older date demands the current view of history be pushed further back into time. Both also suggest that even if the actual stones were laid in place somewhere around those dates, the intelligence behind the project and the knowledge required to maneuver the largest stones and to create the necessary tools to craft them did not arrive "suddenly," but came from an earlier time and people. As we have seen, the global transmission of architectural, mathematical, and engineering skills by the heroes and teachers supports this belief.

Much has been made of the numerous anomalies within the pyramid, and one is of particular interest. The air shafts in the Great Pyramid have been variously explained as simple air shafts, as portals for the pharaoh's soul to escape the tomb after death, or as markers of the precession of the equinox. Controversy rages around the significance, the historical dates, and the stars and constellations that are the targets of those shafts. One constellation that has figured prominently is Orion. Closely identified with Osiris, Orion, according to a number of myths, is where Phaeton was first seen. Found at the foot of Orion is another star associated with the Great Pyramid: Sirius, the Dog Star. Sirius, after remaining below the horizon for 70 days, rose and warned the ancient Egyptians of the annual flooding of the Nile River. Sirius, in Iranian mythology, is associated with Trishtrya the planet killer, and the bringer of the Flood. Sirius the barking dog.

Over the thousands of years that followed the Phaeton disaster, heroes and teachers and their followers preserved as much of the knowledge as they could while rediscovering their world. Quetzalcoatl, Kukulcan, Thoth, and unknown others measured the planet on their journeys. Their skills mapped the new landmasses and rediscovered the dragon paths of energy. Circles of stones were constructed around the world to measure the seasons and the movement of the heavens, as well as the movements of Earth's crust, and to track, as described in *Uriel's Machine,* the courses of comets and warn of future collisions. These people, with their knowledge, wisdom, and projects, brought mankind back from the edge of the abyss.

The myths that tell of Phaeton's devastation of Earth and the great deeds that followed outlived languages, empires and religions, revolutions and inquisitions. Yet there is little doubt that much of the work of the heroes and teachers proved futile. Over the hundreds of years that followed the disaster, there were people who were never visited and others who drove them away. With some who were contacted, each succeeding generation remembered and understood less than the one before; knowledge became muddled, confused, and lost again. So why did our ancient ancestors work so hard to preserve, transmit, and apply this knowledge?

Still strong with the Golden Age spirit, they desired to make a connection to those on the other side of the dark, barbaric age they knew had arrived and to maintain beacons of light for those who would follow them in the darkness. We were here, they wanted to say, before you, and this is how we comprehended the universe. They also wanted to tell us, along with their science, myths, and the stories of this overwhelming disaster, that so much, so many were lost. There is an ancient Middle Eastern tradition that describes the Great Pyramid, located at the heart of the world, as a memorial to that planetary catastrophe and to the memory of all who perished on that day when the world nearly ended.

YESTERDAY, TODAY, AND TOMORROW

Around 1,000 years after Phaeton, Earth's climate underwent its most dramatic change since the disaster. For a number of reasons, including less solar dust in space and less pollution in Earth's atmosphere, Earth began to warm up, and much of the ice created following the disaster began to melt. The Younger Dryas era was ending, and over the course of the next few centuries the seas rose 100 to 300 feet. Settlements of seafaring societies along the coasts were submerged. More than half of Sundaland, which included much of Southeast Asia and Indonesia, sank beneath the rising waters. Across the planet people boarded boats and long ships and sailed away to find new lands or islands. A whole cycle of new myths came into being that, over time, blended with Phaeton's mountain-topping flood legends to produce the hundreds of flood myths we know today. Other coastal people fled up river valleys to the highlands, at times leapfrogging over other races who chose to remain. It is at these higher elevations that scientists have uncovered early evidence of agriculture. As this new environmental disaster spread over the world, humanity remained ignorant of its original cause, but it reconfirmed their vulnerability to the capricious violence of nature and contributed to the psychological trauma begun by Phaeton and passed down to succeeding generations.

Reinforcing this collective shock were side effects from two of the gifts brought by the heroes and teachers. The first gift, domestication, was given to a starving people. In a climatically hostile world, domestication prevented mankind from starving to death. Yet it was a double-edged sword, offering sustenance at a price: Practicing domestication breaks up the world, sculpting a landscape of fences and fields, creating barriers, both physical and mental, between mankind and the natural world. Over the thousands of years that followed, domestication introduced the roots of wealth, greed, and domination, but in the Golden Age there was a balance to its disruptive side. It was the One Law of doing unto others—including *all* other life-forms and not just human. As we have seen, when the heroes and teachers arrived they brought the One Law with them. Some of humanity embraced it and for many thousands of years maintained a more spiritual, balanced view of the world. The goddess societies, which some scholars believe existed 3,000 to 7,000 years ago, were perhaps remnants of those Golden Age traditions.

Others, however, chose not to adhere to the One Law. Some, severely traumatized, could not. The loss of that spiritual anchor unleashed an insidious, devastating reality on the world. Instead of reverence, respect, and participation in the natural world, mankind now lived with detachment and over time worked to control and dominate the planet.

The second gift with side effects that helped alter our ancestor's way of thinking was writing and later its offspring the alphabet. Since the end of the 19th century scientists have been studying the differences between the two sides of the human brain. They have come to understand that the right side of our brain houses the traits we know as holistic and simultaneous, concrete and feminine. On the left side are found those characteristics we call linear and sequential, abstract and masculine. A balanced psyche draws from both sides equally, but when one side of the brain becomes dominant, this upsets the equilibrium and can affect individuals and cultures by creating new unbalanced social patterns and new traumas.

Writing, by its nature, forces the mind into a left-brain pattern of thinking. At first, it began with images whose interpretation is more a right brain function. But as the images became more sophisticated, as with Egyptian hieroglyphs, the process of interpretation moved toward the left side of the brain. As literacy slowly increased over thousands of years, a balanced way of thinking and existing, personified by the Golden Age and the One Law, slowly gave way to one that was unbalanced, focusing more on linear, abstract, and masculine ways of thought and excluding or vilifying right brain ideas, attributes, and values.

Plato, in *Phaedrus,* recognized the dilemma when he had Thamus, the king of all Egypt, warn Thoth of the danger of writing and how it would give people the "appearance of wisdom, not true wisdom." The ancient Egyptians, to the consternation of Egyptologists, did not write down all their knowledge. In India, the *Vedas* also followed this practice by omitting or hiding certain information until a person received the proper instruction, initiations, and disciplines in order to truly understand the wisdom and to prove themselves worthy. Even within the last 100 years tribal peoples, such as the Australian Aborigines and the Dogon of West Africa, have followed this approach—precautionary training.

Is it possible for something that occurred so long ago to

still affect us today? Yes. Since Phaeton, societies and civilizations have continuously retraumatized themselves through child-rearing methods, education, social custom, racism, religion, war, predatory economic policies leading to inequity and poverty, and a separation from nature resulting in the destruction of Earth's environment. It is also known that damaging and unbalanced behaviors can become addictive to individuals and cultures, causing secondary addictions through futile attempts to satisfy needs that current belief systems cannot explain or address. Addictions to power, greed, consumerism, sex, drugs, violence, and abusive behavior are common today in individuals and whole cultures. American ecopsychologist Chellis Glendinning writes, in *My Name Is Chellis & I'm in Recovery from Western Civilization*, "Every trauma that occurs is an individual trauma . . . every trauma is a social trauma with roots in social institutions and implications for society at large, and every trauma is a historic trauma, fostered by the past and reverberating into the future." When trauma is passed down from generation to generation you have a recipe for a cultural disease that can become deeply institutionalized within a civilization.

Yet not every individual or culture is affected in the same way by this ancient disaster and psychological disease. Hunter and gatherer societies, such as the Australian Aborigines and many of the Amerindian cultures, were able to adjust in healthier ways to the trauma of Phaeton and thrived over thousands of years without developing the same symptoms as Western Civilization. Only when they came into contact with "civilized" societies did they begin to exhibit symptoms of the collective disease.

Since Phaeton there have been many forks in the road. On a planet that came close to dying during those terrible days, our ancestors found themselves utterly alone. The heroes and teachers came and went and shared all they had, but they themselves became myths and legends or were forgotten as the centuries passed, and generations of people spread out over Earth like branches of a great tree. As they made their way over land and seas, mankind felt awe in the towering snowcapped mountains and the vast forests that again covered the planet. Formations of migrating birds filled the sky in the spring and the autumn, and the sea was full of fish and great whales. Yet that wonder was now tinged with an instinctive fear of the natural world.

The inability of the original survivors to understand and accept what happened to them and to pass this healing down through the generations left their descendents unprepared and ill-equipped to deal with what lay ahead. Without a guiding force humanity was cast adrift. The most catastrophic fork in the road was the loss of the spiritual connection. The Pygmies of Central Africa believe that this separation from God was the worst disaster to befall our species. In worldwide myth it is called The Fall.

The Fall is the separation of all those after the disaster from those before it. It is the disorientation of being cast out of paradise, losing communion with the spirit, and the pain of not understanding why. To the survivors' descendants, the science behind the "why" was obscured or lost, the peace and plenty before Phaeton were just stories. Whatever explanations were given, they were not enough. Like a small child believing it is he or she who is the cause of the parents' divorce, mankind assumed the role of the guilty party. What exactly we did to deserve such punishment varies with the priorities and imagination of the storyteller, religion, or culture. But generally we became wicked and sinful. We fell.

Since the spiritual connection during the Golden Age was lost, other connections were needed: Mankind created gods to fill the void. Gods of thunder, gods of storms, sky gods from whose hallowed halls fell comets and shooting stars, snow and rain. They were not gods of an age of peace and plenty. They were dangerous gods molded from a world filled with violence, fear, and deprivation. These gods and goddesses took the forms of humans acting and reacting as their human creators did, displaying not only the virtues but the vices of a psychologically dysfunctional species.

What stands out in these new visions is the growing separation between these divine creations and their human creators. These gods and goddesses are remote, making contact with mankind only when they lusted after a mortal or sought revenge for some perceived slight or injury. They are also exclusive. Each existed for a particular people and is found only in a specific part of the world. Over the millennia these gods and goddesses multiplied into polytheistic pantheons and innumerable animistic spirits crowding every bush and tree, every river and mountain. Some held within their personalities the memory of a more holistic interpretation of the spiritual connection. Others did not.

As populations grew, conflicts erupted, flowing from the dark currents of domestication and writing. Consuming addictions to power, greed, and self-interest opened like wounds as our ancestors made their way through the world. Increasingly, patriarchal societies led by dominator kings marched to war and devastated their neighbors, unknowingly imitating Phaeton's ancient blitzkrieg through the solar system. Yet for thousands of years following Phaeton there was universal toleration to all interpretations of the spiritual connection. There were no religious conflicts.

Then the alphabet appeared. According to Leonard Shlain, author of *The Alphabet Versus the Goddess,* "When cultures adopt writing, particularly in its alphabetic form, something negative occurs." He goes on, "An alphabet by definition consists of fewer than thirty meaningless symbols that do not represent the images of anything in particular; a feature that makes them abstract." In order to understand writing constructed with an alphabet, the left side of the brain arranges the individual symbols, letters, into a linear sequence. This process began around 5,000 years ago, 7,000 years after Phaeton's passing. Shlain points out around this time the egalitarian, goddess societies began to disappear at an alarming rate. "Whenever a culture elevates the written word at the expense of the image, patriarchy dominates."

Of the thousands of gods and goddesses that were created by man in Phaeton's wake, one has had the most profound effect on the entire planet. In the beginning known as Yahweh and Jehovah, and later as God the Father and Allah, this interpretation of the spiritual connection took exclusivity and separation to a more menacing level. Here was a new vision of God who shared some of the attributes of his fellow divine inventions: capriciousness, jealousy, humor, and a rage that could become mindless. But he was unique in that he was a singular, all-powerful male god, chaste, frustrated, and lonely as a cloud. He was everywhere. He was abstract. Nowhere was holy. Earth's sacredness was removed from the world and replaced by the written Word of this one new god. He spoke to his special people in a secret language written in an alphabet by the right hand and read in a linear, sequential fashion by the left brain.

These functions combined to push our mind deeper into an unbalanced mental state. Religious conflicts began, and humanity was now forced to seek the spiritual connection outside of itself, to seek it in a book. And in succeeding generations the words in holy books changed subtly in meaning and context as others, with their own agendas, interpreted, edited, or rewrote them. People became separated from the spirit and found themselves mired in a quicksand filled with intolerance and dogma, bigotry and misogyny, hypocrisy and corruption, fear, historical and ecological misconduct, and unfathomable cruelty perpetrated on one another.

Yet the spirit of the Golden Age resurges and flows here and there, now and again, reconnecting with people throughout history. It is found in the hospitality, charity, and compassion of all the world's religions. It is found also in songs and music, like the chords that King David played to sooth Yahweh. When such acts emanate from within a person, from the heart, it is that ancient connection, transcending the written word, seeking expression in the world.

Throughout history, like swells upon the ocean, a few men and women have appeared as great souls, mystics, philosophers, saviors, and visionaries. They succeeded in redirecting the destructive energy of their own unbalanced minds, and reconnected with the Golden Age's universal spirit, still hidden within each of us. They revealed the spirit and the human to be indistinguishable. These new teachers remind us of that Golden Age that today is echoed in the biology and geology of our planet and in our myths and legends, songs and dreams. They remind us of a way of life that has almost been forgotten because 12,000 years ago our ancient ancestors, standing on the edge of the abyss, looked up and saw Phaeton, sublime and monstrous, pass overhead.

FOURTH MEDITATION

The bonfire that blazed through the night, at the assembly, is now a pale fire flickering over dying embers. Dogs lie sleeping at our feet. Darkness is vanishing and a new day is moments away. The cranes, visible at the water's edge, are chattering and dancing. Theseus performed such a dance, in celebration, after killing the Minotaur. The first nine steps are complicated, like finding one's way through a labyrinth, but the tenth is simple: a leap into the sky. The cranes, looking deceptively awkward with their long legs, conclude the first nine steps, and then on the tenth lift off toward the sun, now rising from the horizon.

Thomas Mann wrote, "One can easily live in a story and not understand it." As we have pulled the stick from the water and gazed upon it unbroken, we know this is true. Our current worldview, as depicted by science and religion, is inaccurate. By removing science's denial, frozen in an illusionary Ice Age, and realizing that the Word of God, both comforting and afflicting, is not the source of our spirit, we begin to see clearly. Science and religion are flawed institutions created by unbalanced minds. They are symptoms of a traumatic experience: Phaeton.

Phaeton's passage through our solar system 12,000 years ago was a watershed moment for our species. Regardless of its distance in time, the physical and psychological stress it produced continues to exert a profound influence on our species. To recognize this is the first step to our eventual recovery. Remaining in denial seals our doom.

Ishtar, queen of the Mesopotamian gods, stood before those gods after the Flood and said, "These things I shall not forget, to eternity I shall remember!" We have been on the brink of forgetting. By remembering, we will catch more than a glimpse of paradise. We will step back from the abyss before *us* and take our first steps toward home.

We have reached the open sea, with some charts;
and the firmament.
—DOROTHY DUNNETT

CHAPTER SOURCES AND NOTES

Buddha said it best: "Do not believe anything because the written testimony of some ancient wise man is shown to you. Do not believe anything on the authority either of teachers or priests. Whatever accords with your own experience and after thorough investigation agrees with your reason, and is conducive to your own welfare and to that of all other living things, that accept as truth and live accordingly." Astronomer Fred Hoyle has acknowledged that scientists, "must recognize ourselves for what we are—the priests of a not very popular religion." It is time for a change.

If you are interested in diving deeper into our ancient past, here are some suggestions. These books will continue the journey that began here. What you will discover is that many people have pieces of the great puzzle. Out of this puzzle one day will rise a unified theory about the disaster and its aftermath 12,000 years ago. It will become our new worldview, a new paradigm. You may wish to contribute.

Long before there were links on the Web, or even a Web, there were bibliographies. Within this bibliography are many books that will serve as good points of departure. They will not always agree with one another nor with the contents of this book, but the ideas found in these books all assisted in the writing of *Watermark* and will aid you in your journey as well.

1. IN THE BEGINNING

Allan, D.S., and J.B. Delair, *When the Earth Nearly Died*. Bath: Gateway Books, 1995. Also known as *Cataclysm* in the United States (published by Bear & Company, 2001). Highly recommended. Their dedication in uncovering this ancient mystery was a major inspiration in writing this book, and I have followed their choice in using Phaeton as the name of the invader.

Alverez, Walter, *T. rex and the Crater of Doom*. Princeton, NJ: Princeton University Press, 1997.

Brown, Hugh, *Cataclysms of the Earth*. New York: Twayne Publishers, 1967.

Gould, Stephen, *Dinosaur in the Haystack*. New York: Harmony Books, 1995. See the chapters: "Lucy on the Earth in Stasis" and "Jove's Thunderbolts."

Hapgood, Charles, *Path to the Pole*. New York: Chilton Books, 1970.

Huggett, Richard, *Cataclysms and Earth History*. Oxford: Clarendon Press, 1989.

Raup, David, *The Nemesis Affair*. New York: W. W. Norton, 1986.

de Santillana, Giorgio, and Herta von Dechend, *Hamlet's Mill*. Boston: Nonpareil Books, 1969. A dense and erudite book that astronomer E. C. Krupp calls an "origami nightmare." He may well have a point, but it is also as simple as a lottery. If you have a lottery where the odds are millions to one, and the same person wins that lottery three times in a row, you are going to become suspicious. You are pretty sure that someone has fixed the odds. The myths of the world are entertaining stories, but the recurring themes indicate an intelligence behind them. We may not understand the message completely, but we know a message is there. The myths were fixed.

Velikovsky, Immanuel, *Earth in Upheaval*. London: Abacus, 1973. Velikovsky is an inspiration to anyone who looks at ideas beyond the current paradigm. Though shunned and demonized by the scientific community, many of his ideas are proving to be true. See *Velikovsky Reconsidered* by the editors of *Pensee* to see the true impact of his work.

2. A FANTASY FOR TOMORROW

The end of the world has been described, from the earliest times, in oral history through myths and in written history in apocalyptic literature. The books of Daniel, Enoch, and the Revelation of John represent a style of writing that flourished between 200 BC and AD 200 among the Jews and Christians. Humanity's end was brought about by mankind's wickedness and was punished by a divine being. Today the genre is found primarily in science fiction writing and film. Now the events are either manmade, or mankind is the victim of some greater cosmic event (with a divine being receiving little if any credit). *Armageddon* and *Deep Impact* are two films that explore the possibility of Earth being struck by a comet or an asteroid. Our planet, in both cases, survives (thanks to human bravery, skill, and technology) with only minor damage. An event like Phaeton's visit, however, would be another story entirely.

3. THE TALE OF THE LIVING AND THE DEAD

Allan, D.S., and J.B. Delair, op. cit.

Anonymous, "Cave Yields a 'Noah's Ark' of Ancient Animal Remains." *The Oregonian,* May 23, 1999.

Encyclopedia Britannica, 15th Edition, 1990.

Hooker, Joseph Dalton, *The Botany of the Antarctic Voyage of H.M. Discovery Ships Erebus and Terror, in the Years 1839–1843.* New York: J. Cramer-Weinheim, 1963. Kerguelen Island was once covered by "a luxuriant forest. Throughout many of the lava streams are found prostrate trunks of fossil trees of no mean girth." The lava flows are dated Late Pleistocene, 12,000 years ago.

Kauffmann, Jean-Paul, *The Arch of Kerguelen.* New York: Four Walls Eight Windows, 2000. Literature on the islands of Kerguelen (also known as Desolation Island) in English is slim. It is under French jurisdiction, and research has been published in French. Most information in English comes from either Hooker's field reports or Henry Ridley's work. Kauffmann's book presents a vivid, contemporary account of the islands but touches on their history as well.

Krause, Hans, "The Mammoth and the Flood."
http://hanskrause.de.
A common image of the mammoth is of a hairy elephantlike beast with long tusks in a snow-covered environment or standing below walls of ice on the covers of popular books. Mammoths were about the same size as an African elephant. That is, 10 to 13 feet tall and weighing around 8.5 tons when full grown. They must have required about the same amount of grasses and vegetation as an African elephant to survive, up to 500 pounds per individual per day. They also would have needed upwards of 40 gallons of water per individual per day. In Africa today, elephants can starve to death after four to six months. Their stomachs can be full, but they perish due to the lack of protein and energy.

When you add the hundreds of millions of other animals (and their dietary needs) that lived in the northern regions 12,000 years ago, the orthodox Ice Age melts away.

La Violette, Paul, *Earth Under Fire*. New York: Starlane Publications, 1997.

Ridley, Henry, *The Dispersal of Plants Throughout the World*. Ashford, UK: L Reeve & Co., LTD, 1930.

Stefansson, V., *Greenland*. New York: Doubleday, 1942.

Waller, Geoffrey (editor), *Sealife*. Washington: Smithsonian Institution Press, 1966.

Ward, Peter, *The Call of the Distant Mammoth*. New York: Copernicus, 1997. Ward tells of the dwarf mammoths of Wrangel Island. One hundred miles north of Siberia, this island was the home to mammoths 4,000 years ago. But they were only six feet tall! Their ancestors were survivors of Phaeton who were cut off from the mainland when the landmasses were reconfigured. They had to adapt to a harsher and more restricted environment. They did so by becoming smaller. Evolution happens in small ways.

4. PHAETON, THE SHINING ONE

Allan, D.S., and J.B. Delair, op. cit.

Brakenridge, G.R., "Terrestrial Paleoenvironmental Effects of a Late Quaternary Age Supernova." *Icarus*. April 1981. Vol. 46.

Collective, *The Torah*. New York: Jewish Publishing Society of America, 1962. Psalm 19. Another translation (Jerusalem Bible) reads, "He has his rising on the edge of Heaven," then continues, "the end of his course is it's farthest edge, and nothing can escape his heat."

Encyclopedia Britannica, op. cit.

Gray, L., and John MacCulloch (editors), *The Mythology of All Races*. New York: Cooper Square Publications, 1964.

Heidel, Alexander, *The Babylonian Genesis*. Chicago: University of Chicago Press,1951. The *Enuma Elish*.

Jobes, Gertrude, *Dictionary of Mythology, Folklore and Symbols*. New York: Scarecrow Press, 1962.

Melville, A.D., *Ovid's Metamorphoses*. New York: Oxford University Press, 1987.

O'Byrne, J. (ed.), *Advanced Sky Watching*. New York: Time-Life Books, 1997.

Pliny, *Natural History*. Loeb Classical Library, Vol. II. Boston: Harvard University Press, 1969.

Wilkins, W.J., *Hindu Mythology*. London: Thacker and Company, 1882.

5. EARTH'S STORY

Allan, D.S., and J.B. Delair, op. cit.

Baigent, Michael, *Ancient Traces*. London: Penguin Books, 1999.

Donnelly, Ignatius, *The Destruction of Atlantis*. New York: Multimedia Publishing, 1971.

Encyclopedia Britannica, op. cit.

Feder, Kenneth, *Frauds, Myths, and Mysteries*. Mountain View, CA: Mayfield Publishing, 1990.

Huggett, Richard, op. cit.

Imbrie, John, and Katherine Imbrie, *Ice Ages*. Cambridge, MA: Harvard University Press, 1979.

Krajick, Kevin, "The Riddle of Carolina Bays." *Smithsonian*, September 1997.

La Violette, Paul, op. cit.

Pettersson, Hans, *Westward Ho with the Albatross*. New York: Dutton & Co., 1953.

Schoch, Robert, *Voices of the Rocks*. New York: Harmony Books, 1999.

Velikovsky, Immanuel, op. cit.

6. PHAETON: INTO THE BREACH

Allan, D.S., and J.B. Delair, op. cit.

Anonymous, "Kuiper Belt."
http://solarviews.com/eng/kuiper.htm
http://www.ifa.hawaii.edu/faculty/jewitt/kb/kb.con.html.
Astronomers confirmed in the 1990s that as many as 70,000 objects with diameters larger than 60 miles were moving in orbit around the Sun beyond Pluto. Also found at the rim of the solar system is the Oort Cloud, a vast hypothetical concentration of small bodies. All these objects may be comets or asteroids or both. It is also possible that the Oort and the Kuiper Belt are one and the same: the remains of the tenth planet destroyed by Phaeton 12,000 years ago.

Clube, Victor, and Bill Napier, *The Cosmic Winter*. Oxford, England: Basil Blackwell, 1990.

Davis, John, *Beyond Pluto*. New York: Cambridge University Press, 2001. Davis writes that there is as "much as fifty Earth masses of material" in the Kuiper Belt, or enough for a planet about half the size of Saturn.

Encyclopedia Britannica, op. cit.

Gray, L., and John MacCulloch, op. cit.

Heidel, Alexander, op. cit.

Jobes, Gertrude, op. cit.

Melville, A.D., op. cit.

O'Byrne, J., op. cit.

Pliny, op. cit.

de Santillana, Giorgio, and Herta von Dechend, op. cit.

Temple, Robert, *The Crystal Sun.* London: Arrow Books, 2000. See the chapter, "The Case of the Disappearing Telescope." Temple sums up the denial that runs through much of modern civilization when he says, "people see what they expect to see, and are blind to what they are convinced cannot exist; it is what I call 'consensus blindness.'"

7. OUR ANCIENT ANCESTORS, PART 1

Allan, D.S., and J.B. Delair, op. cit.

Budge, E.A. Wallis, *From Fetish to God in Ancient Egypt.* Oxford: Oxford University Press, 1934.

Cremo, Michael, and Richard Thompson, *Forbidden Archeology.* San Diego, CA: Bhaktivedanta Institute, 1993. Thomas Lee was, unfortunately, not the only scientist to face a scientific inquisition when his discoveries challenged prevailing dogma. The shameful side of science is exposed for all to see.

Deloria,Vine, *Red Earth, White Lies*. Golden, CO: Fulcrum Publishing, 1997.

Denton, Michael, *Evolution: A Theory in Crisis*. Bethesda, MD: Adler & Adler, 1985. Darwin was never totally convinced of his idea. Even in 1872, when the last edition of *Origin* was published, he had many doubts. But by this time, and in the years following, his evolution had "evolved" into dogma, and once a hypothesis becomes a "self-evident truth, its defense becomes irrelevant and there is no longer any point in having to establish its validity by reference to empirical facts."

Dillehay, Thomas, *The Settlement of the Americas: A New Prehistory*. New York: Basic Books, 2000.

Encyclopedia Britannica, op. cit.

Feuerstein, Georg, Subhash Kak, and David Frawley, *In Search of the Cradle of Civilization*. Wheaton, IL: Quest Books, 1997. A portion of human history overlooked by the West is given its due. A few years ago, David Frawley made the observation that much of the imagery in the *Rig Vedas* is oceanic. This supports the view that the Proto-Indo-Europeans, whose descendents would write the *Vedas* and who would be the source of many modern languages, came from the sea. We can only wonder when and from where. We know why: Phaeton.

Flem-Ath, Rand, and Rose Flem-Ath, *When the Sky Fell*. New York: St. Martins Press, 1995.

Hapgood, Charles, *Maps of the Ancient Sea Kings*. Philadelphia: Chilton, 1966. Long, long before Columbus and 1492, humanity sailed the ocean blue.

Knight, Christopher, and Robert Lomas, *Uriel's Machine*. London: Century Books, 1999.

La Violette, Paul, op. cit.

Marshack, Alexander, *The Roots of Civilization*. New York: McGraw Hill, 1972.

Settegast, Mary, *Plato Prehistorian*. Hudson, NY: Lindisfarne Press, 1990.

West, John, *Serpents in the Sky*. Wheaton, IL: Quest Books, 1993.

Myth

Campbell, Joseph, *The Mythic Image*. Princeton, NJ: Princeton University Press, 1981.

———, Forward to Maya Deren, *Divine Horseman: The Living Gods of Haiti*. Syracuse, NY: McPherson, 1981.

Jobes, Gertrude, op. cit.

Lopez, Barry, *Arctic Dreams*. New York: Bantam Books, 1987. See the chapter, "The Country of the Mind."

Luther, Martin, "95 Theses." http://members.aol.com/theclarion/creeds_confessions/luther_95.html

Paradise

Heinberg, Robert, *Memories and Visions of Paradise*. Los Angeles: Jeremy P. Tarcher, 1989.

Melville, A.D., op. cit.

Language

Heinberg, Robert, op. cit.

Rudgley, Richard, *The Lost Civilizations of the Stone Age*. New York: Free Press, 1999. See the chapter "The Mother Tongue."

Health

D'Adamo, Peter, *Eat Right for Your Type*. New York: Putnam's Sons, 1996. See the chapter, "Blood Types."

Deloria, Vine, op. cit.

Encyclopedia Britannica, op. cit.

Lawlor, Robert, *Voices of the First Dawn: Awakening in the Aboriginal Dreamtime*. Rochester, VT: Inner Traditions, 1991.

Noorbergen, Rene, *Secrets of the Lost Races*. New York: Bobbs-Merrill Company, 1978. With good air, healthy food, and long life, how many people were living on Earth 12,000 years ago? Noorbergen suggests before the Flood there were 900 million (the same world population as in the 1800s). At the end of the Younger Dryas era, science estimates there were only 4 million people.

Ward, Peter, op. cit.

Spirit

Fix, William, *Lake of Memory Rising*. San Francisco: Council Oak Books, 1999. The divine spark within us is taken a step further by Fix: We are god. The insight has been acknowledged throughout history. Christ quotes it from Psalm 82 (quoting his father, in fact), "You are gods." Cicero writes, "Know, then, that you are a god." Many find this idea blasphemous, preferring to believe in the separateness of our species, isolated from all things in the universe, including the divine. Yet that insight of divinity has much in common with the more balanced worldview of the Golden Age. As we have seen, most of our myths describe those that lived in that age as godlike in one form or another.

Revised English Bible. Oxford: Oxford and Cambridge University Presses, 1989.

Domestication

Caras, Roger, *A Perfect Harmony*. New York: Simon & Schuster, 1996.

Childress, David, *Lost Cities of Ancient Lumeria and the Pacific*. Stelle, IL: Adventures Unlimited, 1988.

Diamond, Jared, *Guns, Germs and Steel*. New York: W.W. North, 1997.

Encyclopedia Britannica, op. cit.

Isaac, Erich, *Geography of Domestication*. Englewood Cliffs, NJ: Prentice Hall, 1970.

Reynolds, Philip Keep, *The Banana*. New York: Houghton Mifflin, 1927.

Stevens, Henry, *The Recovery of Culture*. New York: Harper, 1949. Stevens suggests the thought-provoking idea that humanity drew inspiration for its columned temples from the magnificent trunks of trees that made up Earth's immense forests.

Medicine

Encyclopedia Britannica, op. cit.

Glausiusz, Josie, "The Iceman Healeth." *Discover*, February 2000.

Noorbergen, Rene, op. cit.

Ronan, Colin, *Lost Discoveries*. New York: McGraw Hill, 1973.

Rossbach, Sarah, *Fen Shui*. New York: Dutton, 1983.

Rudgley, Richard, op. cit. See the chapter, "Stone Age Surgery."

8. PHAETON, THE PLANET KILLER

Sound produced by mechanical waves cannot travel in a vacuum because it requires a medium (air, water, or steel for example) to transport it. This type of sound includes everything you are hearing at this moment. Sounds produced by electromagnetic waves however do not require a medium and can travel through a vacuum. Radio waves are the most obvious example. But radio waves are not within our hearing range. Phaeton's battle with Tiamat, and later it's combat with Mars and Earth, would produce infinitely more powerful electromagnetic waves that would span all the frequencies of sound.

Allan, D.S., and J.B. Delair, op. cit.

Encyclopedia Britannica, op. cit.

Ginenthal, Charles, *Carl Sagan & Immanuel Velikovsky*. Tempe, AZ: New Falcon Publication, 1995. The Tharsis Bulge is a continental sized surface deformity on Mars, 6½ miles high. Volcanoes are common on its slopes, including the gargantuan Olympus Mons, the largest known volcano in the solar system (16 miles high). Branching down from the bulge is a chaotic system of canyons and valleys, sometimes miles deep, including the 2,400 mile long Valles Marineris. Because they are not filled with billions of years of rock and sedimentation, they are considered a recent geologic event. Scientists believe that magma under tremendous pressure came from deep within the planet to form the bulge. Almost directly on the opposite side of Mars is the Hellas Planitia, an impact crater four to six miles deep, and 1,250 miles in diameter. The material thrown out of this crater would cover the continental United States to a depth of two miles. Many scientists assume its age to be about 4 billion years. But not all—Carl Sagan was one who believed it was a recent event. As we have seen, it may well be the impact crater of Tiamat.

Gray, L. and John MacCulloch, op. cit.

Heidel, Alexander, op. cit.

Hines, Daryl, *The Homeric Hymns*. New York: Atheneum, 1972. See "To Ares."

Jobes, Gertrude, op. cit.

McFarling, Usha, "Photos Hint Mars Was Awash in Lakes." *The Oregonian*, December 5, 2000.

Recer, Paul, "Evidence Shows Mars Once Awash." *The Oregonian*, November 30, 2001.

Stutley, Margaret, *Harper Dictionary of Hinduism*. New York: Harper & Row, 1977.

Vergano, Dan, "Rock Basins on Mars Point to Land of Lakes." *USA Today*, December 5, 2000.

9. OUR ANCIENT ANCESTORS, PART 2

Energy

Allan, D.S., and J.B. Delair, op. cit.

Devereux, Paul, *Places of Power*. London: Blandford, 1999.

Lawlor, Robert, op. cit.

Mitchell, John, *The New View Over Atlantis*. New York: Thames & Hudson, 1995.

Morrison, Tony, *Pathways of the Gods*. Lima, Peru: Andean Airmail and Peruvian Times, 1978.

Robins, Don, *Circles of Silence*. London: Souvenir Press, 1985.

Rossbach, Sarah, op. cit.

Thom, Alexander, *Megalithic Sites in Britain*. Oxford: University Press, 1967.

Wichlein, John. "Spirit Paths of the Anasazi." *Archeology,* January/February, 1994.

Architecture

Bauval, Robert, and Adrian Gilbert, *The Orion Mystery*. New York: Crown Trade Paperbacks, 1995.

Collins, Andrew, "Baalbek."
http://andrewcollins.net/page/articles/baalbek.htm

Cornelius, Geoffrey, and Paul Devereux, *The Secret Language of the Stars and Planets*. San Francisco, Chronicle Books, 1996.

Dunn, Christopher, *The Giza Power Plant*. Rochester, VT: Bear & Co., 1998.

Fix, William, *Pyramid Odyssey*. New York: Mayflower, 1978.

Hale, Jonathan, *The Old Way of Seeing*. New York: Houghton Mifflin, 1994.

Hancock, Graham, *Fingerprints of the Gods*. New York: Crown Publishers, 1995.

Hawkins, Gerald, *Stonehenge Decoded*. Garden City, NY: Doubleday, 1965.

———, *Beyond Stonehenge,* New York: Harper & Row, 1973.

Knight, Christopher, and Robert Lomas, op. cit.

Mendelssohn, Kurt, *The Riddles of the Pyramids*. New York: Praeger Publishers, 1975.

Temple, Robert, op. cit. See the chapter, "The Eye of Horus."

Thom, Alexander, op. cit.

Tompkins, Peter, *The Secrets of the Great Pyramids*. New York: Harper & Row, 1978.

Mathematics

Feuerstein, Georg, Subhash Kak, and David Frawley, op. cit.

Hancock, Graham, op. cit.

Hayes, Michael, *The Infinite Harmony*. London: Weidenfeld & Nicholson, 1994.

La Violette, Paul, *Beyond the Big Bang*. Rochester, VT: Park St. Press, 1995.

Lawlor, Robert, *Sacred Geometry*. New York: Thames & Hudson, 1982.

Schwaller de Lubicz, R.A., *Sacred Science*. Rochester, VT: Inner Traditions, 1988.

Tompkins, Peter, *Mysteries of the Mexican Pyramids*. London: Thames & Hudson, 1987.

West, John, op. cit.

Astronomy

Allan, D.S., and J.B. Delair, op. cit.

Bauval, Robert, and Adrian Gilbert, op. cit.

Bellows, Henry, *The Poetic Edda*. New York: Biblo & Tannen, 1969.

Dunn, Christopher, *The Giza Power Plant*, op. cit.

———, "Precision." *Atlantis Rising*, March/April 2002.

Encyclopedia Britannica, op. cit.

Hancock, Graham, op. cit.

Jobes, Gertrude, op. cit.

Krupp, E.C., *Beyond the Blue Horizon*. New York: Harper Collins, 1991.

La Violette, Paul, *Beyond the Big Bang*, op. cit.

Noel, M. and D. Tarling. "Laschamp Geomagnetic Event." *Nature*. February 1975. Vol. 253.

O'Byrne, J. (ed.), op. cit.

Proctor, Richard, *Light Science for Leisure Hours*. London: Longmans, Green And Company, 1873. See chapter: "New Theory of Achilles' Shield."

de Santillana, Giorgio, and Herta von Dechend, op. cit.

Stewart, R.J., *Creation Myths*. Longmead, Dorset: Element Books, 1989.

Sullivan, William, *The Secret of the Incas*. New York: Crown Publishers, 1996.

The Flood

Allan, D.S., and J.B. Delair, op. cit.

Bellamy, H.S., *Moons, Myths and Man*. London: Faber & Faber, 1949.

Bierlein, J.F., *Parallel Myths*. New York: Ballantine Books, 1993.

Bulfinch, Thomas, *Bulfinch's Mythology*. New York: Avenel Books, 1978.

Campbell, Joseph, *The Mythic Image*, op. cit.

Donnelly, Ignatius, *Atlantis: The Antediluvian World*. New York: Gramercy Publishing Company, 1949.

Flem-Ath, Rand, and Rose Flem-Ath, op. cit.

Jobes, Gertrude, op. cit.

MacCulloch, John (ed.), *The Mythology of All Races*. New York: Cooper Square Publications, 1964.

Oppenheimer, Stephen, *Eden in the East*. London: Phoenix, 1998.

de Santillana, Giorgio, and Herta von Dechend, op. cit.

10. PHAETON AND EARTH

Allan, D.S., and J.B. Delair, op. cit.

Anonymous, "Native American Myths of Creation." http://www.crystalinks.com/nativeamcreation.html

Atkinson, Austen, *Earth Impact*. London: Virgin Publishing, 1999.

Bellamy, H.S., op. cit.

Bellows, Henry, op. cit.

Encyclopedia Britannica, op. cit.

Ginenthal, Charles, op. cit.

Gray, L., and John MacCulloch, op. cit.

Jobes, Gertrude, op. cit.

King James Bible, New York: Viking Studio Edition, 1999.

La Violette, Paul, *Earth Under Fire*, op. cit.

Tedlock, Dennis. *Popol Vuh*. New York: Simon and Schuster, 1985.

Wickman, Robert, "Volcanism on the Moon." http://volcano.und.nodak.edu/vwdocs/planet_volcano/lunar/Overview.html

Wilkins, W.J., *Hindu Mythology*. London: Thacker and Company, 1882.

Young, J., *Prose Edda of Snorri*. London: Bowes & Bowes, 1954.

11. AFTER THE FLOOD

The trinity of biology, geology, and our human memory created the case for the disaster brought on by Phaeton's visit, but what of the aftermath? How can we begin to understand what happened to the survivors? Myths reveal some of it, but not all. The enormity of the catastrophe is barely imaginable. Perhaps if we took *every* natural and manmade calamity, from hurricanes and volcanoes, forest fires and tidal waves, earthquakes and collisions with asteroids and comets, famines and plagues to our wars, wars, wars, throughout our recorded history, and have them repeat themselves, completely, all within the next 48 hours, we may get an inkling of the event that our ancient ancestors lived through.

The Heroes

Allan, D.S., and J.B. Delair, op. cit.

Atkinson, Austen, op. cit. There are 16 major meteor streams, including the huge Beta Taurid complex, which rains down on Earth's Northern Hemisphere each year. After 12,000 years the fading remains of Phaeton's armada still shower down on us.

Bellamy, H.S., op. cit.

Bellows, Henry, op. cit.

Boaz, Franz, *Kutena Tales, Bureau of American Ethnology, #59.* Washington, D.C.: Smithsonian Institution, 1918. Franz lists Mount Baker and Mount Ranier in Washington, Mount Jefferson in Oregon, and Mount Shasta in Northern California as sites where "great canoes" beached.

Clube, Victor, and Bill Napier, op. cit. For Clube and Napier, periodicity is the key. Their research confirms that comets and meteors have bombarded Earth periodically for many thousands of years. Groups of narrow growth tree rings and the chemical makeup of Greenland ice cores indicate periods of terrible cold and dense atmospheric dust. These are the bigger and meaner brothers to the annual meteor streams mentioned above.

Gray, L., and John MacCulloch, op. cit.

Heidel, Alexander, op. cit.

Jerusalem Bible, Readers Edition. New York: Doubleday, 1971.

Jobes, Gertrude, op. cit.

King James Bible, op. cit.

La Violette, Paul, *Earth Under Fire,* op. cit.

Tedlock, Dennis, op. cit.

The Teachers

Anonymous, "The Golden Rule."
http://www.fragrant.demon.co.uk/golden.html

——, "The Story of Oannes." http://www.oannes.com/

———,."Quetzalcoatl."
http://weber.ucsd.edu/~anthclub/quetzalcoatl/quetzal.htm

Atkinson, Austen, op. cit.

Bauval, Robert, and Adrian Gilbert, op. cit.

Campbell, Joseph, *The Power of Myth*. New York: Doubleday, 1988.

Clube, Victor, and Bill Napier, op. cit.

Flem-Ath, Rand, and Rose Flem-Ath, op. cit.

Gray, L., and John MacCulloch, op. cit.

Hancock, Graham, op. cit.

Hapgood, Charles, *Maps of the Ancient Sea Kings*, op. cit. The first successful post-Phaeton societies must have been based near the oceans. That they measured the Earth implies navigation and mathematic skills, resulting in the ability to find both latitude and longitude. They did not hug the coasts fearing deep water. They were masters of their craft. The stone circles found around the world, with their mathematical and astronomical precision, can only have been constructed by a people at home upon the sea.

Heyerdahl, Thor, *Early Man & the Ocean*. New York: Vintage, 1980.

Jobes, Gertrude, op. cit.

Knight, Christopher, and Robert Lomas, op. cit.

Lawton, Ian, and Chris Ogilvie-Herald, *Giza, the Truth*. Montpelier, VT: Invisible Cities Press, 2001.

Oppenheimer, Stephen, op. cit.

de Santillana, Giorgio, and Herta von Dechend, op. cit.

Sullivan, William, op. cit.

Tompkins, Peter, *The Secrets of the Great Pyramids*, op. cit.

Yesterday, Today, and Tomorrow

Allen, Charlotte, "The Scholars and the Goddess." *The Atlantic Monthly*, January 2001.

Anonymous, "The Golden Rule," op. cit.

Armstrong, Karen, *A History of God*. New York: Alfred Knopf, 1993.

Barton, George, *Archaeology and the Bible*. Philadelphia: American Sunday School, 1937.

Bloom, Harold, *The Book of J*. New York: Grove Weidenfeld, 1990.

———, *Omens of the Millennium*. New York: Riverhead Books, 1996.

Eisler, Raine, *The Chalice & The Blade*. San Francisco: Harper Collins, 1988.

Encyclopedia Britannica, op. cit.

Fix, William, *Lake of Memory Rising*, op. cit.

Glendinning, Chellis, *My Name Is Chellis and I'm in Recovery from Western Civilization*. Boston: Shambhala, 1994.

Gray, L. and John MacCulloch, op. cit.

Hartmann, Thom, *The Last Hours of Ancient Sunlight*. New York, Harmony Books, 1998.

Heidel, Alexander, op. cit.

Heinberg, Richard, "Catastrophe, Collective Trauma, & the Origin of Civilization." http://www.igc.org/museletter/muse35.html.

Heinberg relates how psychiatrist Robert Lifton concluded, after studying the longterm psychological effects of the Hiroshima attack and the Buffalo Creek, West Virginia, flood in 1972, that "the impact of disasters not only effect immediate victims, but is transmitted to succeeding generations." One has only to consider the chronic conflicts in Northern Ireland, the Balkans, the Middle East, or the misogyny, racism, and religious intolerance that are present in Western society to understand that a monstrous trauma has been living within our civilization for a long time.

Logan, Robert, *The Alphabet Effect*. New York: William Morrow, 1986.

Mann, Thomas, *Joseph and His Brothers.* London: Penguin, 1984.

Noorbergen, Rene, op. cit.

Shlain, Leonard, *The Alphabet Versus the Goddess*. New York: Viking, 1998. How humanity's brain was rewired in the thousands of years that followed the Phaeton disaster. http://www.alphabetvsgoddess.com

Temple, Robert, *The Sirius Mystery*. Rochester, VT: Destiny Books, 1998.

Turnbull, Colin, *The Mountain People*. London: Picador, 1974. In the last half of the 20th century the Ik people, an African tribe of hunters, were relocated from their hunting grounds and asked to become farmers in a land with no rain. In less than five decades all social organization had collapsed, and the family as a cohesive unit no longer existed. Only the strongest and the most cunning survived. Now imagine a post-Phaeton world.

Legends say that Homer's epic the *Iliad* was once written in its entirety on the inside of a nutshell. Both that story and the one told within these covers may seem fantastic, but both have ample evidence to support their validity. Robert Temple's *The Crystal Sun* will satisfy your curiosity about the miniature reproduction of the *Iliad,* except for the type of nut used. For the story told here, work your way through the "links." Each will open new doors of perception and begin an infinite succession of ideas. We will never know the complete story of Phaeton. What *is* important, however, is that the event happened, our ancient ancestors survived, and that it had a profound and traumatic effect on our species. These facts, when they become the worldview, have the power to change our world. If you doubt this, continue the journey begun here and see for yourself.

ACKNOWLEDGMENTS

We all travel long, strange roads in our lives, collecting nourishment and inspiration from springs and wells along the way. I recall seeing George Pal's *When Worlds Collide* when I was young, and it inspired me to seek out and read Edwin Balmer and Philip Wylie's books about that fictional world disaster. Later, after reading Immanuel Velikovsky's *Worlds in Collision* books, I marveled at the questions raised and the many paths of inquiry it offered. These and others became the kindling, and in 1997 I found the match: D. S. Allan and J. B. Delair's *When the Earth Nearly Died*. They pointed the way. To them and all the other heroes and teachers in my life, I offer my sincere thanks.

I am also grateful to Ann Sihler, Karen Jones, Shanna Germain, Pat Vivian, Gina Bacon, and Nancy Woods for the continual encouragement and positive suggestions these excellent writers and Funny Ladies have given me.

Thank you to my mother, Terry, for having the faith, my brother, Stephen, and that travelin' man Mark Throop for encouragement. And to Patrick Huyghe at Paraview for saying yes!

Watermark would not have been written without Beverly Christy-Vitale, who sustains me in all I do and enriches the lives of all who know her. Thank you.

INDEX

acupuncture, 95–96
"Adam's wood," 24
Aesop's fables, 84
after the Flood, 147–78
  cosmic bombardments, 155–56
  devastation, 154–55
  diaspora, 154
  extinction, 157
  heroes, 147–51, 172
  remembering, 178
  starvation, 153–54, 168
  survival, 151–54, 158
  teachers, 157–67, 172
Age of Aquarius, 120–21
age of darkness, 135, 167
Age of Pisces, 120
Alaska:
  bays in, 52
  muck beds in, 25–26
Allah/Jehovah/Yahweh, 175
Allan, Derek, 24–25, 38, 48, 60
alphabet, 169, 174–75
Alps, formation of, 43–44, 56, 142
Aluminium-26, 65
Alvarez, Louis and Walter, 4
Andes mountains, 44–45, 56
  agriculture of, 163
  ley lines in, 110–11
  Machu Picchu in, 124
  mythology of, 162
  Sacsayhuaman ruins, 113
animals:
  after the Flood, 156–57
  bones and fossils of, 27–31, 46, 52
  canine family, 123–24
  communication with, 84
  domestication of, 90–91, 162, 168–69
  extinct, 4, 24–25, 27, 155, 157
  migratory cycles of, 109
anomaly, use of term, 3–4
Antarctica, 55–56, 57
Apache Indians, 127
apocalypse, 7
Appalachia, 16, 20, 46
archeoastronomy, 125
architecture, 112–15, 124
Asteroid Belt, 99
astronomy, 119–25, 165–66

Australia:
  Great Dreamtime in, 85
  hunter-gatherers in, 171
  precautionary training in, 170
  song lines in, 111
Avebury, England, 124
Ayurveda, 95
Azores Islands, 47
Aztecs, 36, 78–79, 127, 153

Baalbek ruins, 112–13
bananas, 92
bat droppings, 95
"bays," 51–52
Bellamy, H., 128
Bengston, John, 83
Beringia land bridge, 77–78
Berossus, 160
Biblical Flood, *see* Flood
Biblical literalism, 52, 53, 63
biology, 13–17
blood types, 85–86
bone-caves, 27–28, 53, 58
botany, 18–20, 93–95
brain, right and left sides, 169–70
Budge, Wallis, 75

cabbage, 19
Cactus Hill, Virginia, 78
calendar, introduction of, 162
Camden, South Carolina, 51
Campbell, Joseph, 80, 81
Canary Islands, 14, 20, 47
carbon dioxide, 86
cardiac window, 94
Carolina Bays, 51
Carter, Howard, 80
Cascade Range, 44
Catal Huyuk, Turkey, 76–77
Catastrophism, 2, 52, 60, 63
caves and fissures, 26–29, 150–51, 153
Charon, 65
Cherokee Indians, 127
Chile-Peru Trench, 46
China:
  ancient medicine in, 95–96
  Imperial Library in, 123
  paths of the dragon, 108–11
Chiron, 70–71
Choukoutien, China, 27–28
civilization:
  birth of, 74–75, 150–51
  early evidence of, 76–77
  one family of, 88
Clarke, Arthur C., 88
Clausiliacea, 17
clay, red, 49–50
climate change, 25, 167–68
Coastal Range, 44
cold, 58–59
Columbia Plateau, 47
comets, 38–39, 70, 155
Comet Shoemaker-Levy, 70
Confucius, 89
Con Ticci Viracocha, 162–63
continental drift, 3, 15, 18, 20, 58
craters, 51
Creationism, 73
creation myths, 88, 154–55
Creek-Natchez Indians, 127
Cumberland, Maryland, 28

Darwin, Charles, 3, 73–74
Darwinism, historical, 74, 76, 77
"Death Star," 8–11
Deccan Plateau, 46–47
Dechend, Herta von, 121, 123, 163
Delair, J. Bernard, 24–25, 38, 48, 60
Desana Indians, 119
Deucalion, 127
diaspora, 154
dinosaurs, extinction of, 4, 37, 38, 63
Diodorus Siculus, 161

disorientation, 172–73
Dogon, 170
domestication, 90–93, 162, 168–69
Donnelly, Ignatius, 59
dragon paths, 108–11, 166
drift deposits, 21–23, 25, 26, 55, 58
driftwood, 24
Druids, 110
drumlins, 53, 56, 57
Dunn, Christopher, 113–14

Earth:
  axis of, 119–21, 133, 137–38, 148, 156
  biology of, 13–17
  botany of, 18–20
  climate changes on, 25, 167–68
  continents of, 147
  cosmic assault on, 49–52, 155–56
  crust of, 48–49, 58, 136, 142, 166
  energy of, 38, 108–11
  flood of, *see* Flood
  geology of, 3, 21–31, 43–49
  maps of, 75–76
  Moon of, 134, 138, 159
  orbit of, 58
  and Phaeton, 99, 100, 102–3, 127, 131–45, 166
  rain on, 140
  red clay of, 49–50
  rotation of, 120, 135
  sacredness of, 175
  size of, 49
  and Sun, *see* Sun
  upheaval of, 43–47
  water on, 152–53
  winter/Ice Age, *see* Ice Age
earthworms, 14–15
Ebers Papyrus, 94–95
Eckhart, Meister, 87
Egypt, 75, 76, 77, 80, 82, 94–95, 117, 129
  early civilization of, 75, 77, 94, 117
  Ebers Papyrus, 94–95
  hieroglyphs, 170
  King Tut's tomb, 80
  mythology, 36, 82, 119–20, 161, 163–65
  pyramids, 113–14, 164–65
Eldredge, Niles, 4
Electra, 101–2
energy, 38, 108–11, 166
England, 25, 27, 124, 125
Enuma Elish, 36, 68, 70, 99, 100
equinox, 120–22, 163
erratics, 53–55, 58
eskers, 53, 56–57
Eskimo myths, 120
Eugenia (tree), 18–19
Euhemerus, 80
evolution, 3, 73–74
extinction, 4, 24–25, 37, 38, 63, 155, 157

Fall, The, 172–76
feng shui, 108
Fennoscandia, 23–24
Fix, William, 88, 89
fjords, 48
Flood, 2, 24, 125–28, 141–43
  after, *see* after the Flood
  argument against, 52–53, 63
  arks, 92, 128, 132, 134, 137, 141, 142–44, 148–49
  date of, 129–30
  hero myths, 147–54
  mountain-topping, 126, 168
  and Phaeton, 68, 131–45
  safe havens sought in, 69–70, 92, 133, 134, 137, 143
  swells from the sea, 126
  warnings of, 68, 126–28

fly, wingless, 17
Four Horsemen of the Apocalypse, 69

Galápagos Islands, 82
Gangetic Trough, 46
geology, 3, 21–31
Germany, 25
Giza Power Plant, The (Dunn), 113–14
glaciers, 56, 58, 156
Glendinning, Chellis, 171
God, Word of, 177–78
goddess societies, 169, 174–75
Golden Age, 92, 129, 163, 173
  astronomy in, 119
  and goddess societies, 169
  One Law of, 158, 160, 169, 170
  spirit of, 158, 160, 167, 176
Golden Post, 122
Golden Rule, 89
Golden Section, 116
Goldstream Valley, Alaska, 23
Gould, Stephen Jay, 4
gravity, 56, 127–28
Great Abyss, 36
Great Han Hai, China, 45
Great Rift Valley, 47–48
Greek myths, 36
Greenberg, Joseph, 83
Greenland, 23, 55, 57

Haida Indians, 79
Hamlet's Mill (Santillana and von Dechend), 121, 123, 163
Hancock, Graham, 123
Hapgood, Charles, 75–76, 164
Hara (mythical destroyer), 36
harmony, 114–15, 117, 118
Hawaiian Islands, 20
Hawkins, Gerald, 125
health, myths of, 84–87
Heard Island, 17, 19
Hermes/Thoth, 164, 166, 170
Heteromeyenia ryderi, 15–16
Himalayan Mountains, 45–46, 56, 142
Hipparchus, 121
Homer, 80
Hooker, J. D., 16–17, 19, 20
Hopi Indians, 79
horticulture, 92–93
Huang Ti, 95
human body, energy in, 109–10
human family, 88
human remains, 28
hunter-gatherers, 171
hurricanes, 61
hypothesis, creation of, 4, 59

ice, 52–59
  causes of, 59
  meltwater of, 57–58, 168
  movement of, 56
  sheets of, 56, 57–58
Ice Age, 3
  contraindications of, 32, 55–56, 57–59, 63–64
  end of, 16, 57, 90
  fossils of, 27–28, 30–31, 46
  hypothesis of, 19, 24, 30, 52–53, 63, 73
  Phaeton and, 144–45
Iceland, 23
Ice Man (Otzi), 96
immortality, 86–87
impact hypothesis, 4
Inca, 113, 127
India, 36, 95, 170
Ishtar, 178

Japan, 36
Jehovah/Allah/Yahweh, 175
Jupiter, 38, 70, 72

Kaboi, 160

Kali Yuga, 123
karma, 89
Kerguelen Island, 16–17, 19
Kingu, 99–100, 101, 104, 132, 137, 138–39, 141
Knight, Christopher, 125
Kotelnyy Island, 24
Kuiper Belt, 66
Kukulcan, 161–62, 166

La Brea Tar Pits, 29
Lagoa do Sumidouro, Brazil, 28
Lake of Memory Rising (Fix), 88
Lake Tanganyika, 48
Lake Titicaca, 44
Lake Zurich, Switzerland, 25
language, 83–84, 91, 169–70, 174–75
Laschamp geomagnetic event, 119
lava flows, 46–47
Lee, Thomas, 78
ley lines, 108–11
lignite, 21–23, 25
Logos, 89
Lomas, Robert, 125
longitude, finding, 76
luminosity, 87
Lyell, Charles, 3, 4, 52, 73

Machu Picchu, 124
magnetic poles, 109
mammals, *see* animals
manganese, 50
Manitoulin Island, 78
Mann, Thomas, 177
maps, 75–76, 164
Marduk (avenging god), 36, 144
markers, 123–24
Marquesas Islands, 162
Mars, 102–6
Marshack, Alexander, 74–75
Maryland, bays of, 51
mathematics, 115–18
Maya people, 79
McDonald island group, 19
medicine, 93–97
meditations:
  first, 31–33
  second, 62–64
  third, 128–30
  fourth, 177–78
Mehrharh, Pakistan, 76–77
meltwater, 57–58
Mesopotamia, 36, 127
Metamorphoses (Ovid), 36, 37, 67
meteors, 155–56
Mexico, pyramids in, 117
Mid-Atlantic Ridge, 47
Minotaur, 177
Miocene Ocean, 16, 45
Mitchell, John, 111
Mojave Indians, 127
monk seals, 15
Monte Verde, Chile, 78
Moon, 134, 138, 159
Moons, Myths and Man (Bellamy), 128
moraines, 53, 55–56, 58
Mount Pinatubo, 61
Mount St. Helens, 61
muck beds, 26, 29
Musa paradisiacal banana, 92
myths, 79–90
  of astronomy, 119–25
  birth of humanity, 150–51
  caves in, 150–51
  celestial cycle, 121
  creation, 88
  of destruction, 36–37
  of flood, *see* Flood
  health, 84–87
  hero, 147–51, 161, 172
  historical interpretation of, 80
  language of, 83–84
  of luminosity, 87
  markers of, 123–24

myths *(cont.)*
multidimensional nature of, 80–81
of Paradise, 81–83, 86, 87
remote gods of, 173–74
and spirit, 88–90, 158
teachers of, 157–67, 172

Najas flexilis, 18
Natural History (Pliny), 66
Neolithic Age, 74
Neptune, 38, 66–68, 71
New Siberian Islands, 24
New Stone Age, 74
New Zealand, 19, 48
nickel, 49–50
nirvana, 89
Noah, 68, 126, 127, 148
"Noah's wood," 24
Nordenskiöld, A. E., 75
numbers, 115–18, 122–24, 163

Oannes, 161
Occam's Razor, 59
oceanic islands, 20
oligochete, 14–15
one family, 88
One Law, 89, 158, 160, 169, 170
open-heart surgery, 94
Origin of Species (Darwin), 73–74
Orion, 36, 166
Osiris, 161, 163, 166
Otzi (Ice Man), 96
Ovid, 36, 37, 67

Pacific Ocean:
mythology of, 162–63
red clay of, 49
sea bottom of, 46
Paradise, 81–83, 86, 87
permafrost, 156
Persian myths, 120
Phaeton, 37–41, 99–106, 168
birth of, 65
chariot of, 69
and Earth, 99, 100, 102–3, 127, 131–45, 166
and Jupiter, 72
and Mars, 102–6
and Neptune, 66–68
and Saturn, 70–71, 72
and the Sun, 62, 145, 146, 155
and Tiamat, 99–103
and Uranus, 68–69
and Vela, 39–40
and Venus, 145–46
and Wormwood, 152
planets, 35, 37–39
plants:
domestication of, 92
medicinal value of, 93–97
plate tectonics, 3, 48
Plato, 129, 170
Pleiades, 36
Pleistocene Age, 53, 90; *see also* Ice Age
Pliny the Elder, 66
Pluto, 65, 67
Point Barrow, Alaska, 52
Polaris, 120
poles, wandering of, 58
polyhedron, 48
Popol Vuh, 154
Portolans (maps), 75–76, 164
post-traumatic stress disorder, 157–58
prairie mounds, 53, 56, 57, 58
precession, 120–22, 124, 163
prime meridians, 76
Principles of Geology (Lyell), 3
Proctor, Richard, 119
Prometheus, 127, 161
proportion, 116–17
Proto-Global language, 83
Ptolemy, 76

"punctuated equilibrium" hypothesis, 4
pyramids, 113–14, 117, 164–65, 167

Queen Charlotte Islands, 79
Queen Elizabeth Islands, 23
Quetzalcoatl, 161–62, 166

rabbit myths, 150–51
reincarnation, 86–87
religion:
  separatism of, 109, 175–76
  and spirit, 88–90, 176
  Word of God, 177–78
remembering, 178
revelation, 89
Roche's Law, 37–38
Rocky Mountains, 44, 56, 142
Roman roads, 110
Roots of Civilization (Marshack), 74–75
Ruhlen, Merrit, 83

Sacsayhuaman ruins, 113
Santa Paula Mountains, 26
de Santillana, Giorgio, 121, 123, 163
Satan, 36
Saturn, 38, 70–71, 72
scale, 60–62
Schliemann, Heinrich, 80
Set, 36
SETI (Search for Extraterrestrial Intelligence), 118
Seven Devils Canyon, 47
"sewage pharmacology," 95
Shiva, 36
Shlain, Leonard, 174–75
Siberia, 24, 29–30, 55, 58
Sierra Nevada Mountains, 44
Sirius (Dog Star), 166
Smith, Edwin, 95
Snake River, 47
South Georgia Island, 19
spirit, 88–90, 158, 176
Spitsbergen, 23, 26, 55
sponges, 15–16
star, life of, 61
Stone Age, 165
stone circles, 111, 166
Stonehenge, 125
striations, 55
subglacial hydrology, 58
Sun:
  eclipse of, 72
  and equinox, 120
  and Phaeton, 62, 145, 146, 155
  return of, 158–59
Sundaland, 168
supernova, 8, 39–41, 61
Susa-no-wo, 36

tachylite, 47
teal, 19
Tenochitlán, 78–79
Teotihuacán, 124
Tezcatlipoca, 36
Thom, Alexander, 125
Thoth/Hermes, 164, 166, 170
Tiamat, 99–103, 104, 141, 152
Tibetan Plateau, 45
Tiki, 162
Tishtrya, 36–37
Tlaloc, 127
Tlandrokpah, Chief, 150
trephining/trepanning, 93–94
Tristan de Cunha, 19–20
Triton, 67
Tut, King, 80
Typhon (monster), 36

Uniformitarianism/Uniformity, 3, 4–5, 52, 54, 60–61, 62, 63, 73
United States, ley lines in, 110

universe:
  origin of, 88
  spiritual, 88–89
Uranus, 38, 68–69, 71
Uriel's Machine (Knight and Lomas), 125, 166
Utnapishtim (Noah), 68, 127, 147, 148

Valhalla, fighters of, 123
Vallonet cave, 27
Vedas, 95, 116, 170
Vega, 120
vegetarianism, 84–86, 153
Vela, 39–40
Velikovsky, Immanuel, 4, 80
Venus, 145–46
volcanic ash, 49
volcanism, 46–47, 61, 134, 149, 156

water:
  symbolism of, 153
  undrinkable, 152–53
Watkins, Alfred, 108
West, John, 75
When the Earth Nearly Died (Allan and Delair), 24–25
William of Occam, 59
Word of God, 89
Worlds in Collision (Velikovsky), 4
Wormwood, 152–53

Yahweh/Jehovah/Allah, 175
Younger Dryas, 154, 168

Zuni Indians, 148